SPICEY
CIRCUITS
Elements of Computer-Aided
Circuit Analysis

CRC Press
Computer Engineering Series

Series Editor
Udo W. Pooch
Texas A&M University

Published books:

Telecommunications and Networking
Udo W. Pooch, Texas A&M University
Denis P. Machuel, Telecommunications Consultant
John T. McCahn, Networking Consultant

Spicey Circuits: Elements of Computer-Aided Circuit Analysis
Rahul Chattergy, University of Hawaii

Forthcoming books:

Discrete Event Simulation: A Practical Approach
Udo W. Pooch, Texas A&M University
James A. Wall, Simulation Consultant

Microprocessor-Based Parallel Architecture for Reliable Digital Signal Processing Systems
Alan D. George, Florida State University
Lois Wright Hawkes, Florida State University

Algorithms for Computer Arithmetic
Alan Parker, Georgia Institute of Technology

Handbook of Software Engineering
Udo W. Pooch, Texas A&M University

SPICEY CIRCUITS

Elements of Computer-Aided Circuit Analysis

Rahul Chattergy

Department of Electrical Engineering
University of Hawaii
Honolulu, Hawaii

CRC Press

Boca Raton   Ann Arbor   London   Tokyo

Library of Congress Cataloging-in-Publication Data is available from the Library of Congress.

This book represents information obtained from authentic and highly regarded sources. Reprinted material is quoted with permission, and sources are indicated. A wide variety of references are listed. Every reasonable effort has been made to give reliable data and information, but the authors and the publisher cannot assume responsibility for the validity of all materials or for the consequences of their use.

Direct all inquiries to CRC Press, Inc., 2000 Corporate Blvd., N. W., Boca Raton, Florida, 33431.

©1992 by CRC Press, Inc.

International Standard Book Number 0-8493-7173-2

Printed in the United States 1 2 3 4 5 6 7 8 9 0
Printed on acid-free paper

TABLE OF CONTENTS

Chapter 1 Fundamentals

Chapter 2 Resistive Circuits

Chapter 3 Introduction to SPICE

Chapter 4 Controlled Sources

Chapter 5 Linear Circuit Theorem

Chapter 6 Circuit Equations

Chapter 7 Energy-Storage Elements

Chapter 8 Natural Responses of RC and RL Circuits

Chapter 9 Step-Responses of RC and RL Circuits

Chapter 10 Responses of RLC Circuits

Chapter 11 Sinusoidal Analysis

Chapter 12 Frequency Response

Appendix A Algebra of Complex Numbers

PREFACE

A goal of every electrical engineer is to use circuit analysis to understand circuit behavior. However, circuit behavior is an abstract concept and what is considered to be an acceptable behavior for a particular circuit changes from one instance to the next. In one instance we may be interested in the behavior of the output voltage as the frequency of the input voltage is changed. In another we may want to know how the value of a feed-back resistor affects the output voltage. In any case, no matter how circuit behavior is defined in a particular instance, it is obvious that one set of numerical values of circuit variables is not enough to describe the behavior of a circuit.

Circuit analysis is a type of problem solving. There are well known methods of circuit analysis. However, a thorough understanding of each individual method is not enough to gain competence in circuit analysis. It can not be done by a mechanical application of techniques. One must develop some judgement or feel for circuits.

In teaching circuit analysis, the commonly used approach is to derive general results and then to reinforce the concepts by means of examples. That is fine as far as it goes; however, what gets reinforced depends on what is stressed in the examples. Numerical examples, i.e., number crunching for isolated cases, are virtually useless for providing insights into circuit behavior.

As an example consider current division. Once the formulas have been derived, isolated numerical examples at best help us memorize these formulas. But they do not clarify their significance. This later goal is easier to attain if we analyze the behavior of the component currents with variations in parameter values and relate this to the circuit structure. The principal tool of circuit analysis is then calculus and not number crunching.

We have not included a large number of numerical problems at the end of each chapter because we found such problems stress the wrong notions of circuit analysis. We have not provided answers either. Rather, we have discussed techniques of assessing the validity of expressions obtained by simple circuit manipulations.

Complex expressions derived in circuit analysis are often difficult to interpret by manual analysis. It is here that computer aided circuit analysis becomes helpful. The computer and its power of number crunching is used not to solve isolated cases but to generate families of solutions or graphs by simulation. Such graphs bring out the meaning hidden behind a complex algebraic expression. To obtain maximum benefit from the aid of a computer one has to carefully weave computer simulation techniques into analysis itself and not postpone it until the end. Today, this is much easier to do because of the wide availability of personal computers.

We have attempted to write a text on elementary circuit analysis while weaving in computer simulation techniques in the analysis itself. The simulation program used is the famous SPICE (Simulation Program with Integrated Circuit Emphasis) which appears to have become an industry standard. Our goals are (1) to understand circuit behaviors by means of computer simulations, and (2) to learn to use SPICE at an introductory level.

Although there are other simulation packages available, SPICE is not difficult to use at the introductory level. Since SPICE is widely used in the industry, knowledge of how to use this program can only benefit an engineer. The labor will be worth it if the text helps beginners in circuit analysis appreciate how computers can be used as a beneficial tool rather than a hindrance to understanding.

We have been very conservative in our selection of topics for this text. Currently, introductory texts in circuits run anywhere from seven hundred to a thousand pages. Many of the topics included such as three-phase circuits, op-amp circuits, magnetic circuits, Fourier and Laplace transforms are cursorily covered. As Rohrer points out ("Taking circuits seriously," Circuits and Devices, July, 1990, pp. 27-31) "much of the introductory circuits course has come to be preparatory to subsequent courses." We have selected only those topics that are absolutely fundamental to circuit analysis namely, resistor circuits with constant-output sources, energy storage elements and transients and sinusoidal response. The subsequent courses in power, control, and electronics can include their own preparatory materials.

Special thanks are due to Professors Richard Halverson and Vinod Malhotra for many constructive criticisms, and to Professor Udo W. Pooch for constant encouragement.

Rahul Chattergy
University of Hawaii
Honolulu

This book is dedicated in gratitude and love to

Lyda Opura Tanco

the other woman in our family.
She would, if she could, and she did when we needed her most.
Thanks Lyda.

1

FUNDAMENTALS

Static electricity generated by friction had been known to exist at least since the time of the ancient Greeks. It was observed experimentally that two electrified objects either attracted or repelled each other. To explain the existence of two distinct forces it was necessary to assume the existence of two distinct types of electrical charges. These are currently known as the *positive* charge and the *negative* charge. Experiments indicate that opposite types of charges attract and similar types repel. Precise measurements of electrical forces were carried out by Coulomb leading to his fundamental law.

1.1 Coulomb's Law

According to Coulomb's law, the magnitude of the force $\mathbf{F}$ (measured in Newtons) between two small spherical, electrically charged objects separated by a distance $\mathbf{d}$ (measured in meters) in vacuum and carrying charges $\mathbf{q}_1$ and $\mathbf{q}_2$ is given by

$$\mathbf{F} = \mathbf{k}\, \mathbf{q}_1\, \mathbf{q}_2 \,/\, \mathbf{d}^2,$$

where $\mathbf{k} = 9 \times 10^{+9}$. Coulomb's law can be used to create units for measuring electrical charge. Consider a situation where $\mathbf{q}_1 = \mathbf{q}_2 = \mathbf{q}$ and $\mathbf{d} = 1$. Using Coulomb's law we can write

$$q = \sqrt{F/k}\ .$$

Using this equation $\mathbf{q}$ is by definition one unit when $\mathbf{F} = \mathbf{k}$. This unit of charge is called a *coulomb*. Note that for $\mathbf{F}$ to equal $\mathbf{k}$, it has to be of the order of two billion pounds. Hence, one coulomb is an enormous amount of electrical charge. An electron, the smallest unit of negative electrical charge, has a charge of 1.6×10^{-19} coulombs.

1.2 Current

Electrical current is caused by electrical charge in motion. We can define *current* as the *rate* of flow of electrical charge in the direction normal to a unit area. The unit of current is *ampere* and by definition one ampere equals one coulomb per second.

Example 1.2.1

Current in an electrical conductor is produced by the drift of free electrons acted upon by an externally applied electrical force. Let us compute the drift velocity of electrons in a copper conductor of one centimeter square cross section and one meter length carrying a current of one ampere.

Using the following definitions

A = cross sectional area,
L = length,
V = volume,
N = number of free electrons per unit volume,
Q = charge,
I = current,
T = time, and
e = charge of an electron,

we can write,

$$V = A \, L,$$
$$Q = e \, N \, V = e \, N \, A \, L,$$
$$I = Q \, / \, T = e \, N \, A \, L \, / \, T.$$
$$drift \; velocity = L \, / \, T \; = I \, / \, (e \, N \, A).$$

For copper $N = 8.54 \times 10^{28}$ per m³, and substituting the values given we obtain drift velocity = 0.73×10^{-6} meters per second. Clearly the average electron motion is extremely slow and yet it is sufficient to maintain appreciable current in the conductor.

1.3 Voltage

Electrically charged objects produce electrical fields in their immediate environment. Movement of electrical charge in an electrical field produced by some other charge causes change of energy of the system. To relate the quantity of charge moved to the amount of change in energy level would require knowledge of the laws governing the strength of an electrical field. To avoid these difficulties we shall make use of a model of an analogous situation.

We assume the reader is familiar with the law of gravitation and the concept of *potential energy* in a gravitational field. The potential energy $\mathbf{P}$ of an object of mass $\mathbf{m}$ at a height $\mathbf{x}$ above the surface of the earth is given by $\mathbf{P} = \mathbf{m}\, \mathbf{g}\, \mathbf{x}$, where $\mathbf{g}$ is the constant of gravitation. Since $\mathbf{g}$ is independent of $\mathbf{x}$ we can introduce a new concept called the *gravitational potential* at a point $\mathbf{x}$ and define it to be $\mathbf{h_x} = \mathbf{g}\, \mathbf{x}$. In terms of this potential $\mathbf{h_x}$, the potential energy is given by $\mathbf{P} = \mathbf{m}\, \mathbf{h_x}$. Following this analogy we assume that every point in an electrical field has an electrical potential denoted by $\mathbf{v_x}$, and the change of energy $\mathbf{E_{xy}}$ (measured in *joules*) caused by the movement of charge $\mathbf{q}$ from $\mathbf{x}$ to $\mathbf{y}$ in the field is given by $\mathbf{E_{xy}} = \mathbf{q}\, (\mathbf{v_x} - \mathbf{v_y})$. The *potential difference* $\mathbf{v_x} - \mathbf{v_y}$ is measured in a unit called *volts*. Frequently the difference $\mathbf{v_x} - \mathbf{v_y}$ is called the *voltage* between $\mathbf{x}$ and $\mathbf{y}$. An incremental definition of voltage is: $\mathbf{v} = d\mathbf{E}/d\mathbf{q}$. As with gravitational potential we assume that the electrical potential of the ground is zero.

1.4 Power

Power is defined as the rate of change of energy with time. Let $\mathbf{E}(t)$, and $\mathbf{P}(t)$ denote the energy and power respectively at time t. Then by definition $\mathbf{P}(t) = d\mathbf{E}(t)/dt$. In an electrical system we have

$$\begin{aligned}
\mathbf{P}(t) \; &= d\mathbf{E}(t)/dt, \\
&= d\mathbf{E}/d\mathbf{q} \; \; d\mathbf{q}/dt, \\
&= \mathbf{v}(t)\, \mathbf{i}(t).
\end{aligned}$$

Power can be measured in units of joules per second or in volt-ampere. This unit is given the name *watts*.

1.5   Circuit Element

An electrical circuit is constructed out of *circuit elements*. We can think of a circuit element as a box with a finite number of terminals as shown in Figure 1.1 (a). A circuit is constructed by joining the terminals of circuit elements into a structure (consult Figure 1.1 (b)). Electrical charge flows in or out of each terminal of every circuit element in a circuit. Hence, each terminal of a circuit element is marked with the symbol of a current. Also, between any pair of terminals, there is a voltage (consult Figure 1.1 (c)). The relationships among the terminal voltages and currents of a circuit element depends on the physical nature of the element. These terminal relationships or equations are all that are necessary for us to know in order to analyze an electrical circuit.

Figure 1.1 (a). Circuit element.

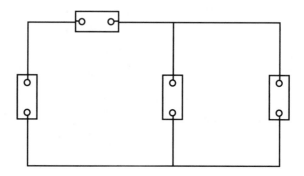

Figure 1.1 (b). Circuit.

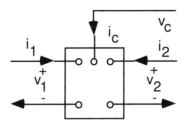

Figure 1.1 (c). Terminal currents and voltages.

1.6 Sign Convention

Voltage is the difference of electrical potential between two points in an electrical field. Being a difference it can be either positive or negative, i.e., it has an associated sign. Similarly, current being a rate of flow can also be either positive or negative. Hence, voltages and currents are *algebraic* quantities, i.e., they have associated signs. To analyze a circuit we need to write mathematical equations involving voltages and currents. To obtain the correct equations we need to assign a consistent set of signs to the associated voltages and currents. The sign convention discussed below serves this purpose.

Passive sign convention is the most commonly used sign convention for assigning signs to voltages and currents. A *passive* circuit element does not generate energy. Figure 1.2 (a) shows a two terminal passive circuit element. According to our sign convention the current is considered positive when it points into the terminal marked +ve for the voltage. Note that the sign of the terminal voltage is related to that of the terminal current and only one of them is assigned arbitrarily. The other one is determined with respect to the one already assigned. Hence, the element shown in Figure 1.2 (b) is also marked according to the passive sign convention.

Since voltages and currents are algebraic quantities, so is power. Under passive sign convention if power computed for an element is positive, then that element is absorbing energy. Otherwise it is delivering energy.

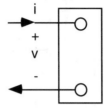

Figure 1.2 (a). Two terminal passive sign convention.

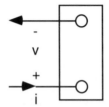

Figure 1.2 (b). Two terminal passive sign convention again.

1.7 Circuit Concepts

An electrical circuit is constructed by joining circuit elements at the terminals. Figure 1.3 shows a typical circuit. Wherever two or more elements are joined, a *node* is created. The nodes in Figure 1.3 are numbered and shown by black circles. Any element stretching between two nodes forms a *branch* of the circuit. A sequence of branches joining two nodes forms a *path* between the two nodes. In general, for a given pair of nodes it is possible to have multiple paths joining them. Clearly, a path can also be thought of as a sequence of nodes by picking the nodes at the end of each branch in the path as intermediate nodes in the path. A path has a *loop* in it if an intermediate node is repeated. A *closed path* is any path where the starting node is identical to the ending node. For our use we shall consider closed paths without loops. Such closed paths are often called loops.

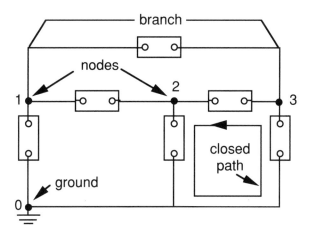

Figure 1.3. Typical circuit.

1.8 Active Sources

An *active* source in a circuit is a source of energy used to maintain a voltage or a current at a certain node. The interaction of all the active sources in a circuit set up the currents and voltages in the rest of the circuit. Henceforth we shall refer to them as sources.

A source is primarily characterized by its terminal output. The output of a *voltage source* is a voltage and that of a *current source* is a current. If the output of a voltage source is independent of its terminal current, it is called an *ideal* voltage source. An ideal current source is defined in a similar manner. If the output of a source is independent of time it is called a *constant* source. Frequently it is also called a *dc* source. Figure 1.4 (a) shows the graphic symbols for ideal, constant, voltage and current sources. Their terminal voltage/current relationships are shown in Figure 1.4 (b).

Figure 1.4 (a). Ideal constant sources.

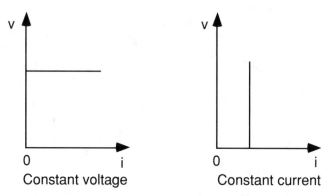

Figure 1.4 (b). Terminal voltage/current relations.

The terminal output of a *controlled* source is controlled by voltages and/or currents from other parts of a circuit. The control equation can be either linear or nonlinear. Linear controlled sources are essential for modelling electronic circuits. Nonlinear controlled sources have their uses too. These sources will be discussed in detail in a later chapter. Figure 1.4 (c) shows the graphic symbols used to represent ideal controlled sources.

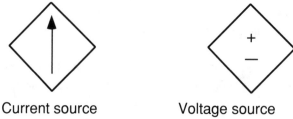

Figure 1.4 (c). Ideal controlled sources.

1.9 Interpretation of Graphs

In later chapters we shall make use of graphs to shed light on relationships given by algebraic formulas. The nature of a graph can tell us a lot about such a relationship at a glance. For illustration let us consider a relationship between a system parameter and the system's response whose graph is shown in Figure 1.5 (a).

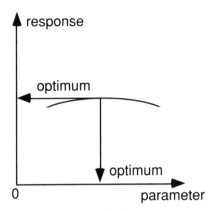

Figure 1.5 (a). Hypothetical system response.

If the problem is to set the parameter value to optimum in order to maintain an optimum response, then the curve in Figure 1.5 (a) indicates a satisfactory relationship. If the parameter value strays from the optimum, the system response does not greatly degrade. However, if the problem is to use the optimum system response to find an unknown optimum value of a system parameter, then this curve indicates a most unsatisfactory relationship. Large deviations in parameter value causes only small changes in system response. Next, suppose this relationship is as graphed in Figure 1.5 (b).

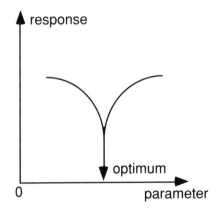

Figure 1.5 (b). Hypothetical system response.

This curve shows an excellent opportunity for finding the optimum value of a system parameter from the system response. However, the system response is extremely sensitive to the parameter value which must be carefully controlled to maintain optimum system response. Finally, consider the curves graphed in Figure 1.5 (c).

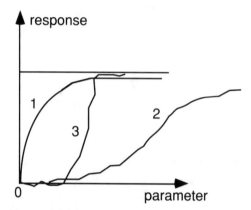

Figure 1.5 (c). Hypothetical system responses

Curves 1 and 2 show that the system responses asymptotically approach a constant as the parameter value increases. In case of curve 1 the initial rate of increase is much higher than that in case of curve 2. On the other hand, curve 3 shows a completely different situation. In this case the system response appears to jump from a low value to a high value as the parameter value passes over some critical interval. Intuitively one realizes that this indicates some sharp change in the state of the system, and such changes are not possible in passive linear circuits. Hence, such response characteristics are only to be expected for nonlinear circuits.

1.10 Summary

• Coulomb's law gives the force between two electrically charged bodies.

• Current is the rate of flow of charge in the direction normal to a unit area.

• Voltage is the potential difference between two points in an electrical field.

- Power measures the rate of change of energy with time.

- Circuit elements have multiple terminals. Knowledge of their terminal voltage/current relations are necessary for circuit analysis.

- Passive sign convention is used to obtain a correct and consistent set of mathematical equations describing a circuit.

- Nodes, branches, paths, closed paths and loops are useful concepts related to a circuit. See Section 1.7 for details.

- Sources are assumed to be ideal. There are independent as well as controlled sources. See Section 1.8 for details.

2

RESISTIVE CIRCUITS

A purely resistive circuit is built out of circuit elements called resistors and voltage and/or current sources. The sources are assumed to be ideal independent sources with constant terminal outputs. The basic circuit laws, which are valid for all linear circuits, are introduced in this simple setting.

2.1 Resistors

In Chapter 1 we mentioned that circuit elements are characterized by the relationships among their terminal voltages and currents. A resistor is a simple circuit element with only two terminals. Figure 2.1 (a) shows the graphic symbol used in this text for a *resistor* along with its terminal voltage and current. The relationship between v_T and i_T given by functions such as

$$v_T = f(\, i_T \,), \qquad\qquad (2.1.1)$$
$$i_T = g(\, v_T \,), \qquad\qquad (2.1.2)$$

describe the nature of a resistor. For a *linear* resistor, the functions f and g are linear and the above equations take the forms

$$v_T = \mathbf{R}\ i_T, \qquad\qquad (2.1.3)$$
$$i_T = \mathbf{G}\ v_T, \qquad\qquad (2.1.4)$$

where $\mathbf{R} = 1/\mathbf{G}$ is called the *resistance* and $\mathbf{G} = 1/\mathbf{R}$ is called the *conductance* of the resistor. This linear relationship between v_T and i_T is known as *Ohm's law*. Figure 2.1 (b) shows a graphical representation of Ohm's law for three different linear resistors.

13

Since the terminal voltage/current relation of a linear resistor follows Ohm's law, the power in a resistor can be written as

$$\mathbf{p} = \mathbf{v_T}\,\mathbf{i_T} \tag{2.1.5}$$
$$= (\mathbf{v_T})^2\,/\,\mathbf{R} \tag{2.1.6}$$
$$= (\mathbf{i_T})^2\,\mathbf{R}. \tag{2.1.7}$$

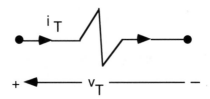

Figure 2.1 (a). Symbol of a resistor.

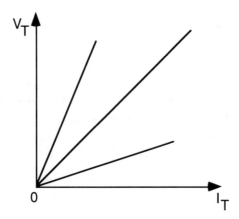

Figure 2.1 (b). Ohm's law.

2.2 Circuit Laws

The voltages and currents in a circuit satisfy *Kirchhoff's laws*. In Chapter 1 we have introduced the concepts of a node, a path, and a closed path in a circuit. In terms of these concepts, Kirchhoff's laws can be stated as follows:

Kirchhoff's voltage law: The algebraic sum of the branch voltages around any closed path in a circuit is zero.

Kirchhoff's current law: The algebraic sum of the currents entering any node of a circuit is zero.

Ohm's law and Kirchhoff's laws can be directly used to find the equivalent resistances of series and parallel combinations of resistors. They also lead to the very useful voltage and current division formulas.

2.3 Resistors in Series

Figure 2.2 shows two linear resistors in series. Using Kirchhoff's voltage and current laws we have

$$\mathbf{v_T} = \mathbf{v_1} + \mathbf{v_2}, \tag{2.3.1}$$

$$\mathbf{i_T} = \mathbf{i_1} = \mathbf{i_2}. \tag{2.3.2}$$

From Ohm's law (2.1.3) we get

$$\mathbf{v_1} = \mathbf{R_1}\,\mathbf{i_1}, \tag{2.3.3}$$

$$\mathbf{v_2} = \mathbf{R_2}\,\mathbf{i_2}. \tag{2.3.4}$$

From the above equations we derive by simple algebra

$$\mathbf{v_T}\,/\,\mathbf{i_T} = \mathbf{R_1} + \mathbf{R_2}. \tag{2.3.5}$$

Hence, we can replace a series connection of two resistors $\mathbf{R}_1$ and $\mathbf{R}_2$ by a single equivalent resistor $\mathbf{R}_{eq} = \mathbf{R}_1 + \mathbf{R}_2$. The process can obviously be extended to any finite number of resistors in series.

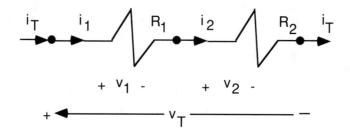

Figure 2.2. Resistors in series.

2.4 Resistors in Parallel

Figure 2.3 shows two linear resistors in parallel. Using Kirchhoff's current and voltage laws we have

$$\mathbf{i}_T = \mathbf{i}_1 + \mathbf{i}_2, \tag{2.4.1}$$
$$\mathbf{v}_T = \mathbf{v}_1 = \mathbf{v}_2. \tag{2.4.2}$$

From Ohm's law (2.1.3) we get

$$\mathbf{v}_1 = \mathbf{R}_1 \, \mathbf{i}_1, \tag{2.4.3}$$
$$\mathbf{v}_2 = \mathbf{R}_2 \, \mathbf{i}_2. \tag{2.4.4}$$

From the above equations we derive by simple algebra

$$\mathbf{v}_T / \mathbf{i}_T = (\mathbf{R}_1 \, \mathbf{R}_2) / (\mathbf{R}_1 + \mathbf{R}_2). \tag{2.4.5}$$

Hence, we can replace a parallel connection of two resistors $\mathbf{R}_1$ and $\mathbf{R}_2$ by a single equivalent resistor $\mathbf{R}_{eq} = (\mathbf{R}_1 \, \mathbf{R}_2) / (\mathbf{R}_1 + \mathbf{R}_2)$. The process can obviously be extended to any finite number of resistors in parallel. However, for more than two resistors in parallel it is more convenient to use conductances and the formula becomes

$$G_q = G_1 + G_2 + G_3 + G_4 + \ldots \qquad (2.4.6)$$

Without loss of generality let us assume that min(R_1, R_2) = R_2. Since R_{eq} = R_2 / (1 + R_2 / R_1), we conclude $R_{eq} \le$ min(R_1, R_2).

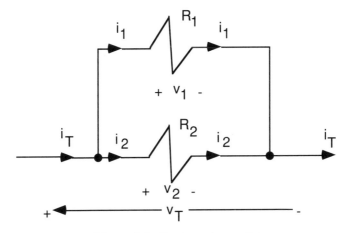

Figure 2.3. Resistors in parallel.

2.5 Current Division

Using the expression of the equivalent resistance R_{eq} for a parallel combination of two resistors R_1 and R_2 (2.4.5) and Ohm's law (2.1.3) we can write (consult Figure 2.3)

$$v_T = i_T \ (R_1 \ R_2) / (R_1 + R_2), \qquad (2.5.1)$$

and since $i_1 = v_T / R_1$ and $i_2 = v_T / R_2$ we also have

$$i_1 = i_T \ R_2 / (R_1 + R_2), \qquad (2.5.2)$$
$$i_2 = i_T \ R_1 / (R_1 + R_2). \qquad (2.5.3)$$

The two factors multiplying i_T are fractions, and because of Kirchhoff's current law these fractions add up to one. Hence, these formulas for i_1 and i_2 show how the terminal current i_T is divided between the two resistors R_1 and R_2 in parallel. Let us test these formulas to see if they agree with our intuitive notion of current division.

If R_1 is very small compared to R_2, then we expect most of the current to be in R_1. Hence, as R_1 approaches zero, i_1 should approach i_T. Similarly as R_1 approaches infinity, i.e., becomes very large compared to R_2, i_1 should approach zero and i_2 should approach i_T. Also when R_1 equals R_2, i_1 should equal i_2. By direct inspection we see that the above formulas satisfy these conditions.

Figure 2.4 shows graphs of i_1 and i_2 as functions of R_1. In this plot, $i_T = 4$ mA and $R_2 = 4K$ ohms. When R_1 is zero i_1 is 4 mA and when R_1 equals R_2 which is 4K ohms, i_1 equals i_2 equals 2 mA. As the value of R_1 increases, i_1 decreases and i_2 increases in value. Also the two current values always add up to 4 mA. Hence, these graphs show us at a glance what current division means. These graphs were obtained by using SPICE and serve as examples of how SPICE can be used to gain insight into otherwise abstract algebraic formulas.

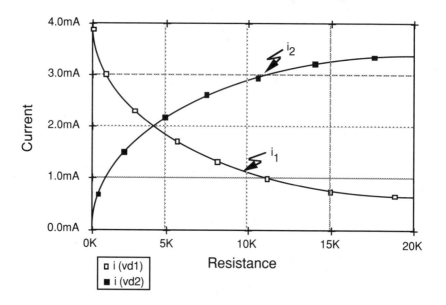

Figure 2.4. Current division.

2.6 Voltage Division

Using (2.3.5) the expression of the equivalent resistance $\mathbf{R}_{eq}$ for a series combination of two resistors $\mathbf{R}_1$ and $\mathbf{R}_2$ and Ohm's law (2.1.3) we can write (consult Figure 2.2)

$$\mathbf{i}_T = \mathbf{v}_T / (\mathbf{R}_1 + \mathbf{R}_2), \qquad\qquad (2.6.1)$$

and since $\mathbf{v}_1 = \mathbf{i}_T \mathbf{R}_1$ and $\mathbf{v}_2 = \mathbf{i}_T \mathbf{R}_2$ we also have

$$\mathbf{v}_1 = \mathbf{v}_T \mathbf{R}_1 / (\mathbf{R}_1 + \mathbf{R}_2), \qquad\qquad (2.6.2)$$
$$\mathbf{v}_2 = \mathbf{v}_T \mathbf{R}_2 / (\mathbf{R}_1 + \mathbf{R}_2), \qquad\qquad (2.6.3a)$$
$$= \mathbf{v}_T / (1 + \mathbf{R}_1 / \mathbf{R}_2). \qquad\qquad (2.6.3b)$$

Again the two factors multiplying $\mathbf{v}_T$ are fractions, and because of Kirchhoff's voltage law they add up to one. Hence, these formulas for $\mathbf{v}_1$ and $\mathbf{v}_2$ show how the terminal voltage $\mathbf{v}_T$ is divided between the two resistors $\mathbf{R}_1$ and $\mathbf{R}_2$ in series. These formulas can be analyzed as was done in the previous section and are left to the reader for that purpose.

2.7 Loaded Voltage Divider

The current and voltage distributions in a simple voltage divider change considerably when a load resistance is placed in parallel with one of the resistors. Figure 2.5 shows a voltage divider with a load resistor $\mathbf{R}$ in parallel with the resistor $\mathbf{R}_2$.

Intuitively we can see that the resistor $\mathbf{R}$ will draw some current from the source $\mathbf{v}_T$. This will increase the current in the resistor $\mathbf{R}_1$ and by Ohm's law, the voltage drop across $\mathbf{R}_1$ will increase. Assuming that the voltage output of the source remains constant, by Kirchhoff's voltage law the voltage across $\mathbf{R}_2$ will decrease. This situation can be verified by analysis.

The equivalent resistance of the parallel combination of $\mathbf{R}_2$ and $\mathbf{R}$ is $\mathbf{R}_{eq} = \mathbf{R} \, \mathbf{R}_2 /$ $(\mathbf{R} + \mathbf{R}_2)$. Using the equations (2.6.3 a and b) for $\mathbf{v}_2$ we obtain the voltage drop across $\mathbf{R}$ as

$$v_R = v_T \, \mathbf{R}_{eq} / (\, \mathbf{R}_1 + \mathbf{R}_{eq} \,)$$
$$= v_T / (\, f(\mathbf{R}) + (\, \mathbf{R}_1 / \mathbf{R}_2 \,)), \qquad\qquad (2.7.1)$$

where $f(\mathbf{R}) = 1 + (\, \mathbf{R}_1 / \mathbf{R} \,)$. Since the resistance values are nonnegative $f(\mathbf{R}) \geq 1$ and hence $v_R \leq v_2$ given by (2.6.3). As the value of $\mathbf{R}$ increases, it draws less current from the source and v_R approaches v_2 from below. In the limit when $\mathbf{R}$ becomes infinite it draws no current at all and the circuit becomes a simple voltage divider. Then the values of v_1 and v_R are given by the formulas of voltage division (2.6.2 and 2.6.3). As the value of $\mathbf{R}$ decreases, it draws more current from the source and increases v_1. Since v_T is constant, and $v_R + v_1 = v_T$ by Kirchhoff's voltage law, v_R approaches zero. More of the source voltage appears across $\mathbf{R}_1$.

The current through $\mathbf{R}_1$ is $v_T / (\, \mathbf{R}_1 + \mathbf{R}_{eq} \,)$ and since $\mathbf{R}_{eq} \leq \mathbf{R}_2$, this current obviously is greater than i_T (2.6.1). Figure 2.6, again obtained by using SPICE, shows how v_1 and v_R changes as a function of $\mathbf{R}$. In this graph $v_T = 10$ V, $\mathbf{R}_1 = \mathbf{R}_2 = 4K$ ohms. As the value of $\mathbf{R}$ increases from 0, the value of v_R given by $v(2,0)$ increases from 0 and approaches 5 V which is the value dictated by the simple voltage division formula in this case. The sum of the two voltage drops always equals the source voltage output of 10 V.

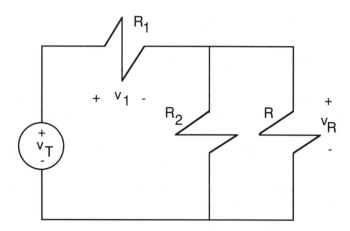

Figure 2.5. Loaded voltage resistor.

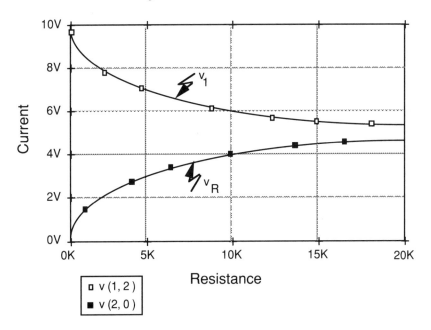

Figure 2.6. Voltage division.

2.8 Wheatstone Bridge

A Wheatstone bridge circuit shown in Figure 2.7 is widely used in various types of
sensitive electrical measuring instruments and transducers. The horizontal branch
completing the bridge has a current measuring device called an ammeter. The bridge is
said to be *balanced* when the ammeter reading is zero, i.e., there is no current in the
horizontal branch. The relationship among the values of the four resistances in a balanced
bridge can be derived as shown below.

In a balanced bridge, since the current in the horizontal branch is zero, the voltages at
nodes 2 and 3 must be the same. Also, under balanced condition, the two arms of the
bridge act as simple voltage dividers. Hence, $v_R = v_3$ and $v_S - v_R = v_S - v_3$.
Furthermore, by voltage division $v_3 = v_S R_3 / (R_1 + R_3)$ and $v_R = v_S R / (R_2 + R)$.
By simple algebra we obtain

$$\mathbf{R}_1\,\mathbf{R} = \mathbf{R}_2\,\mathbf{R}_3, \tag{2.8.1}$$

$$\mathbf{R} = (\,\mathbf{R}_2\,/\,\mathbf{R}_1\,)\,\mathbf{R}_3. \tag{2.8.2}$$

If $\mathbf{R}$ is an unknown resistance, its value can be experimentally obtained by using a balanced Wheatstone bridge and the above equation. The ratio $\mathbf{R}_2\,/\,\mathbf{R}_1$ is changed by factors of 10 followed by variations in the value of $\mathbf{R}_3$ until balance condition is reached. In later chapters this circuit will be analyzed in more detail.

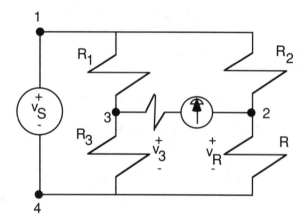

Figure 2.7. Wheatstone bridge.

2.9 Examples

Example 2.9.1

Find the equivalent resistance between nodes 1 and 4 in Figure 2.8.

Using parallel combination of resistances, the equivalent resistance between nodes 1 and 2 is $\mathbf{R}/2$. This in series with the resistance $\mathbf{R}/2$ between nodes 2 and 4 gives a total resistance of $\mathbf{R}$ along the upper path from node 1 to node 4. By series combination, the equivalent resistance between nodes 1 and 4 along the lower path through node 3 is R. Since there are two parallel paths between nodes 1 and 4, by parallel combination the equivalent resistance between these nodes is $\mathbf{R}/2$.

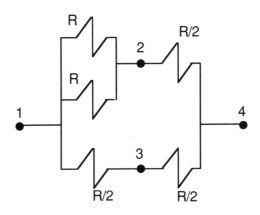

Figure 2.8. Equivalent Resistance.

Example 2.9.2

In Figure 2.9, **R** is a variable resistor. Find the value of **R** that causes the power in **R** to attain its maximum value.

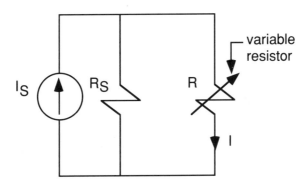

Figure 2.9. Variable Resistance.

By current division, the current in **R** is $\mathbf{I} = \mathbf{I_S}\,\mathbf{R_S}\,/\,(\mathbf{R_S} + \mathbf{R})$. Since power **p** in **R** is given by $\mathbf{I}^2\,\mathbf{R}$ we have

$$p = I_S^2 R_S^2 R / (R_S + R)^2,$$

and

$$dp/dR = I_S^2 R_S^2 / (R_S + R)^2 - 2 I_S^2 R_S^2 R / (R_S + R)^3.$$

Setting dp/dR to zero and solving for R we have $R = R_S$.

Example 2.9.3

Figure 2.10 shows a *delta* connection of three resistors R_a, R_b, R_c, among three nodes a, b, and c. Find the equivalent resistances between pairs of nodes R_{ab}, R_{bc}, and R_{ca}.

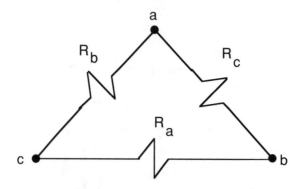

Figure 2.10. Delta resistor connection.

Between nodes a and b we have two parallel paths. The equivalent resistance of the path through node c is $R_b + R_a$. The direct path between nodes a and b has a resistance of R_c. Hence,

$$R_{ab} = R_c (R_b + R_a) / (R_c + R_b + R_a).$$

Expressions for R_{bc} and R_{ca} can be obtained in a similar manner.

Example 2.9.4

Figure 2.11 shows a *star* connection of three resistors R_1, R_2, R_3 among three nodes a, b, and c. Find expressions for R_1, R_2, and R_3 such that the equivalent resistances between pairs of nodes are the same as in Example 2.9.3.

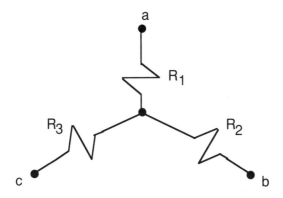

Figure 2.11. Star resistor connection.

The equivalent resistance in the star connection between nodes a and b is $R_{ab} = R_1 + R_2$. This must equal the expression for R_{ab} found in Example 2.9.3. Hence,

$$R_1 + R_2 = R_c (R_b + R_a) / (R_c + R_b + R_a).$$

Similar equations can be derived for $R_2 + R_3$ and $R_3 + R_1$. The required expressions for R_1, R_2, and R_3 can be obtained by solving these equations. These simple derivations are left as an exercise to the reader.

Example 2.9.5

Figure 2.12 shows two alternate ways of connecting four 60 watts, 110 volts light bulbs across a 220 volts source. Discuss the advantages and the disadvantages of the two alternate connections.

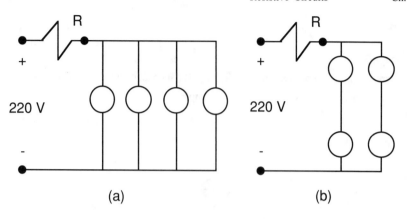

Figure 2.12. Alternate resistor connection.

The connection shown in part (a) of Figure 2.12 will fail more gracefully than that shown in part (b) in case of bulb burnouts. It is possible in part (b) for only two bulb burnouts to cause complete darkness.

Since the bulbs are rated at 110V we must use a voltage divider to drop the excess 220 - 110 = 110 volts of the source across a resistor **R** as shown in Figure 2.12. In part (a), the total current drawn by all four bulbs from the source equal $I = 4$ (60 W / 110 V) = 2.1818 A. Hence, **R** = 110 V / 2.1818 A = 50.4171 ohms. The power in **R** is $p = I^2 R$ = 240 W.

In part (b), two bulbs in series rate 110 V + 110 V = 220 V and can be connected directly across the source. Hence, **R** = 0 and there is no power in **R**. It follows that the connection shown in part (b) wastes less energy than that shown in part (a).

Note that in part (a), if a bulb burns out, then the value of **I** is decreased. This causes a decrease in the voltage drop across **R** according to Ohm's law and increases the voltage across the remaining bulbs. This surely leads to reduced life expectancy for the remaining bulbs. The situation gets worse for the remaining bulbs as more bulbs start to burn out.

Example 2.9.6

Find expressions for the voltage v_0 and the current i_0 in the two circuits shown in Figure 2.13.

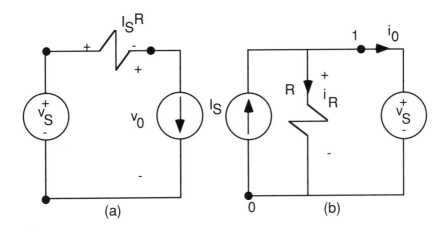

Figure 2.13. Independent constant voltage source.

The current in the circuit shown in part (a) of the Figure 2.13 is $\mathbf{I_S}$ since the source is an independent constant current source. By Ohm's law the voltage drop across **R** is $\mathbf{I_S}$ **R**, and the sign of this voltage drop shown in Figure 2.13 (a) follows the passive sign convention. Using Kirchhoff's voltage law around the only loop in the circuit and rearranging terms we have $\mathbf{v_0} = \mathbf{V_S} - \mathbf{I_S}\,\mathbf{R}$.

In part (b) of Figure 2.13, the voltage   across **R** is $\mathbf{V_S}$ since the source is an independent constant voltage source. By Ohm's law $\mathbf{i_R} = \mathbf{V_S}\,/\,\mathbf{R}$, and using Kirchhoff's current law at node 1 we have $-\mathbf{I_S} + \mathbf{i_R} + \mathbf{i_0} = 0$. Hence, $\mathbf{i_0} = \mathbf{I_S} - \mathbf{V_S}\,/\,\mathbf{R}$.

Example 2.9.7

Find an expression for the current i_3 in Figure 2.14.

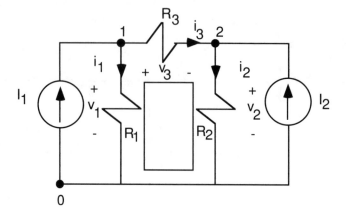

Figure 2.14. Dual voltage sources.

By Ohm's law $v_3 = i_3 R_3$ and $v_1 = i_1 R_1 = (I_1 - i_3) R_1$. Using Kirchhoff's voltage law around the closed path shown in Figure 2.14 we have $- v_1 + v_3 + v_2 = 0$. From this equation we can derive an expression for v_2 as a function of i_3 and it is $v_2 = I_1 R_1 - (R_1 + R_3) i_3$. By Ohm's law $i_2 = v_2 / R_2$. Using Kirchhoff's current law at node 2 in Figure 2.14 we have $I_2 + i_3 - i_2 = 0$. Now substituting v_2 / R_2 for i_2 in this equation and solving for i_3 we have

$$i_3 = (I_1 R_1 - I_2 R_2) / (R_1 + R_2 + R_3).$$

2.10 Validation of Expressions

So far we have shown examples of how to derive expressions for circuit variables by using the circuit laws and the terminal voltage/current relations of the circuit elements. However, how do we know that our derivations are flawless and that these expressions are correct?

This problem is similar to that of ascertaining if a computer program is correct or bug-free. For a computer program, a partial solution to this problem is testing, although testing can not gurantee that a program is correct. However, a successful test increases our confidence in the correctness of a program. Hence, to check on the correctness of an expression we should also test it.

An expression for a circuit variable depends on various circuit parameters. To test it we follow the steps given below.

1. Select parameter values that results in a simple circuit. For resistors we can use values zero and infinity. The outputs of sources can also be set to zeros as needed.

2. Analyze the simple circuit and obtain an expression for the circuit variable under test.

3. Use the parameter values selected in step 1 in the original expression under test and check to see if it reduces to the expression obtained in step 2. If it does, then the expression under test is very likely to be correct.

Step 1 provides a major boost to our understanding of circuits. To select parameter values to simplify a circuit requires reflection beyond that needed to numerically evaluate an expression. One starts to see how the various parts of a circuit are interacting and why a particular expression is the way it is.

Let us consider Example 2.9.7 for illustration. The expression derived for i_3 shows contributions from the two sources I_1 and I_2.

$$i_3 = (I_1 R_1 - I_2 R_2) / (R_1 + R_2 + R_3).$$

With knowledge of the principle of superposition (discussed in a later chapter) one can see at a glance that the expression for i_3 is correct. However, even without such knowledge a lot can be said about this expression.

From Figure 2.14 we see that the source I_1 by itself will send a current through R_3 in the direction of i_3. Similarly source I_2 by itself will produce a current in R_3 in the direction *opposite* from i_3. Thus the minus sign in the numerator of the expression for i_3 is to be expected.

If R_2 is set to zero, then all the current from I_2 will go through R_2 since this will be the path of least resistance across I_2. The circuit now becomes simple and i_3 can be obtained by simple current division to be

$$i_3 = I_1 \, R_1 \, / \, (R_1 + R_3).$$

As can be easily verified, this same expression can be obtained by setting $R_2 = 0$ in the original expression for i_3. A similar test can be performed by setting $R_1 = 0$.

Next we consider the limiting case as R_2 approaches infinity. As the value of R_2 increases, the current through it decreases and ultimately all the current from the source I_2 will go through R_3 in a direction opposite of i_3. Also since R_2 is increasing the current through R_2 contributed by the source I_1 will decrease and will ultimately go to zero. The corresponding expression for i_3 becomes

$$i_3 = - I_2.$$

Again this same expression for i_3 can be obtained by taking the limit as R_2 approaches infinity in the original expression for i_3. So far we have analyzed the original expression for i_3 in two different ways and each time it has been found to be consistent with the changed circuit. Hence, it is very likely to be the correct expression. Note the amount of insight gained into the structure of the circuit shown in Figure 2.14 and the behavior of i_3 in the process.

Figure 2.15 graphically shows the variation of i_3 (i(vd3) in the graph) with R_2. The graph was obtained by using SPICE. Here $I_1 = 4$ mA, $I_2 = 8$ mA, and $R_1 = R_2 = 4K$ ohms. When R_2 is zero, i_3 is 2 mA which comes from source I_1 via. a simple current division. As R_2 approaches infinity, i_3 approaches -8 mA coming from source I_2. Note that the sign of i_3 changes as predicted by the analysis.

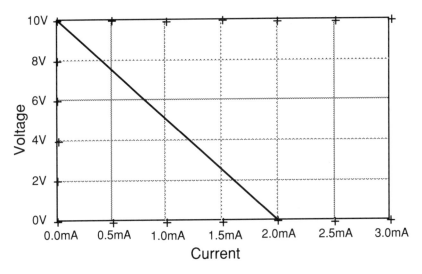

Figure 2.15. Variation of current to resistance.

2.11 Summary

- Ohm's law gives the terminal voltage/current relation of a linear resistor.

- Kirchhoff's circuit laws are needed to write mathematical equations describing a circuit.

- Equivalent resistance of two resistors in series is the sum of their resistance values.

- Equivalent conductance of two resistors in parallel is the sum of their conductance values.

- Voltage (current) division formula is useful for finding expressions of terminal voltages (currents) of two resistors in series (parallel).

2.12   Problems

For every problem, test your derived expressions by the method discussed in Section 2.10.

2.1.   Find an expression for v_3 in the circuit shown in Figure 2.16.

2.2.   Find an expression for v_4 in the circuit shown in Figure 2.17.

2.3.   Find the equivalent resistance of the circuit shown in Figure 2.18.

2.4.   Find an expression for I in the circuit shown in Figure 2.19.

2.5.   Find an expression for I in the circuit shown in Figure 2.20.

2.6.   Find expressions for I in the circuit shown in Figure 2.21 when the variable resistance R_3 has the values zero and infinity.

2.7.   The value of the variable resistance R ranges from 0.5 R_2 to R_2 in the circuit shown in Figure 2.22. Find the range of values for v_R.

2.8.   The value of the variable resistance R ranges from 0.5 R_2 to R_2 in the circuit shown in Figure 2.23. Find the range of values for i_R.

2.9.   In the circuit shown in Figure 2.22, find an expression for R that produces the maximum power in R.

2.10. In the circuit shown in Figure 2.23, find an expression for R that produces the maximum power in R.

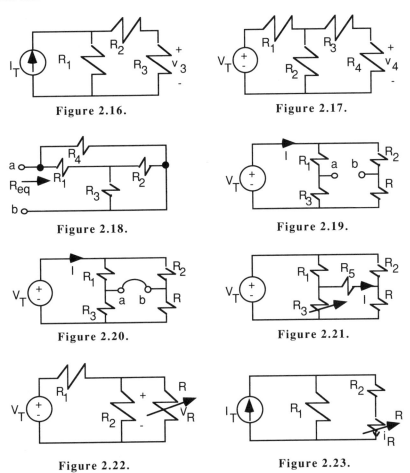

Figure 2.16.

Figure 2.17.

Figure 2.18.

Figure 2.19.

Figure 2.20.

Figure 2.21.

Figure 2.22.

Figure 2.23.

3

INTRODUCTION TO SPICE

The description of the structure of a circuit to be simulated by SPICE is entered in a *circuit file*. A circuit file also contains other information necessary for SPICE to simulate a given circuit. In this chapter we discuss the structure of a circuit file. Any editor that can be used to write computer programs (pure ASCII editors) can also be used to create a circuit file. The name of a circuit file has the extension **.cir**.

3.1 Circuit File

In some SPICE systems a circuit file can be written either in upper or in lower case characters. Only upper case is allowed in some others. Check the documentation of your system to determine the case. We shall use lower case in all our examples.

The first line of a circuit file is the *title* line. It may contain any type of text. The purpose of a title is to give some clue as to the nature of the circuit in the file. It should start with a short descriptive name. Separated from the name by a blank should be a more descriptive text string. Consult Figure 3.1 for a sample circuit file of a simple voltage divider. Presence of blanks in a line is not significant except in the title line.

The last line of a circuit file must have only the string **.end** in it. *Comment* lines start with an "*" in the first column. Comments should explain the purpose of the preceding or the following lines and must not describe what is obvious from the code itself. *Continuation* lines are marked by a "+" in the first column. The lines in between the title and the last need not be in any specific order except in the case of subcircuit definitions. However, it is obviously a good idea to follow some order dictated by the structure of the circuit itself and common sense.

```
vdiv.cir   simple voltage divider
*
*consult Figure 3.2
*element lines start here. vconst is dc
*input source. r1 & r2 are 60 & 40
*ohm resistors in series with source.
*
vconst      1      0       dc      10V
r1          1      2       60
r2          2      0       40
*
*to avoid wasting paper we use
.options nopage
*
*last line in a circuit file
.end
```

Figure 3.1.   Voltage divider circuit.

3.2 Element Line

The *element* lines in a circuit file describe the structure of the circuit to be analyzed as well as introduce the circuit elements used to construct it. To create these element lines, the nodes of a circuit must first be labelled. In some SPICE systems, only nonnegative integers are allowed as labels of nodes. Others allow labels to be alphanumeric strings. Consult Figure 3.2 for an example of a simple voltage divider circuit with labelled nodes.

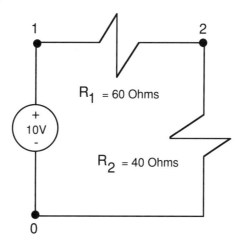

Figure 3.2. Simple voltage divider.

3.2.1 Passive Circuit Elements

Circuit elements in a circuit file are identified by names. A name must start with a letter and the names of resistors, inductors, and capacitors must start with the letters r, l, and c respectively. The number of characters in a name can be as high as 131 however, 8 should be sufficient for most practical purposes. The circuit file in Figure 3.1 has two element lines describing two resistors. These are

$$
\begin{array}{cccc}
\text{r1} & 1 & 2 & 60 \\
\text{r2} & 2 & 0 & 40.
\end{array}
$$

These lines are interpreted as

element   first node   second node   value.

The first element is a resistor named **r1** which connects node 1 to node 2. The value of **r1** is 60 ohms. The second element is another resistor named **r2** and it connects node 2 to node 0. Its value is 40 ohms. Consult circuit diagram in Figure 3.2.

There are many other passive and active circuit elements that can be used to construct circuits. All these can be described in SPICE notations in circuit files. These will be introduced as needed in later chapters. For our current need we shall describe independent sources in a later section.

3.2.2 Element Values

For SPICE simulation of a circuit, values must be assigned to the circuit elements. These values can be written in standard integer, fixed-point, or floating-point notation. Here are some examples.

    5   3.0   -2.71   7E3   -5E-2   2.13E-3

Furthermore, values can also be written in standard scientific or engineering suffix notation. These are

F (femto)	10^{-15}
P (pico)	10^{-12}
N (nano)	10^{-9}
U (micro)	10^{-6}
M (milli)	10^{-3}
K (Kilo)	10^{+3}
MEG (Mega)	10^{+6}
G (Giga)	10^{+9}
T (Tera)	10^{+12}

3.3 Independent Sources

Independent sources are used to drive circuits. In an ideal independent voltage (current) source, the terminal voltage (current) is independent of the terminal current (voltage). The name of an independent voltage source starts with the letter **v**. Similarly the name of a current source starts with the letter **i**. A prefix of **dc** in the value field of a source indicates that its terminal output is constant. Other possibilities are **ac** for a sinusoidal output and a few more time-varying outputs for transient analysis. For now, we shall describe only dc sources.

According to SPICE conventions, the current through a voltage source is positive when it flows from the positive output terminal to the negative output terminal. If power, i.e., the product of terminal voltage and terminal current is negative, then the source is delivering energy to the circuit.

A voltage source with zero output voltage is a *dead* source. The current through such a source as computed by SPICE, is the current in that branch in which the dead source is placed. Hence, such voltage sources can be used as ammeters to find current values in the branches of a circuit. The element lines defining voltage and current sources are shown below.

v dc	n+	n-	dc	xxxV
i dc	m-	m+	dc	yyyA

n+ is the node to which the positive end of **vdc** is connected. m+ is the node through which the current from **idc** leaves the source. The label "dc" in the value field of an element line is optional. However, it is good programming practice to introduce it. xxx and yyy are obviously the output values. The trailing characters V and A are ignored by SPICE but are useful for human comprehension.

3.4 Example

As a simple example of the application of SPICE, let us consider a voltage divider with a load resistance across its output. Consult Figure 3.3 (a) for the circuit diagram. Note that dead voltage sources have been inserted to compute the currents in the two forty ohms resistors. To create the circuit file for this circuit, we follow the steps given below.

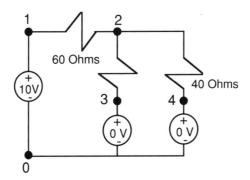

Figure 3.3 (a). Voltage divider with load resistor.

1. Label the nodes in the circuit. The nodes are labelled here with nonnegative integers from 0 to 4. Some SPICE systems allow alphanumeric node labels.
2. Using a pure ASCII editor create the first line of the circuit file as its title line. Consult Figure 3.3 (b).
3. Insert one element line to the file for each element in the circuit: element name, first node, second node, value.
4. Insert the last line (**.end**) to the circuit file.

Note that this circuit file (Figure 3.3 (b)) does not contain any instruction to SPICE as to the nature of the analysis to be performed. SPICE will automatically perform small signal bias point calculation which corresponds to the usual dc analysis.

```
lvdiv.cir    loaded voltage divider
*
*   consult Figure 3.3 (a)

*the same voltage divider as in Figure 3.1
*
vconst     1        0        dc      10V
r1         1        2        60
r2         2        3        40
*r3 is the load resistor in parallel
*with r2 of the voltage divider
*
r3              2        4        40
*
*dead voltage sources v2 & v3 to
*measure currents in r2 & r3
*
v2             3        0        dc      0V
v3             4        0        dc      0V
*
.options nopage
.end
```

Figure 3.3 (b). SPICE circuit file.

The SPICE software is now activated with **lvdiv.cir** as its input file. The sequence of operations needed here changes from one installation to the next. It is best to find out the operation sequence from your own installation. The output of the SPICE simulation is shown in Figure 3.3 (c).

The SPICE output in Figure 3.3 (c) shows the voltages at the nodes with respect to the reference node which in this case is node 0. It also shows the currents through the different voltage sources. Since the sources **v2** and **v3** are dead sources, i.e., with output voltages zero, the currents through these sources are the currents through the resistors in series with them. SPICE also computes the total power dissipation in the circuit elements.

```
lvdiv.cir   loaded voltage divider

****   CIRCUIT DESCRIPTION

*******************************************************
*
*   consult Figure 3.3 (a)
*
*the same voltage divider as in Figure 3.1
*
vconst      1     0     dc      10V
r1          1     2     60
r2          2     3     40
*
*r3 is the load resistor in parallel
*with r2 of the voltage divider
*
r3          2     4     40
*
*dead voltage sources v2 & v3 to
*measure currents in r2 & r3
*
v2          3     0     dc      0V
v3          4     0     dc      0V
*
.options nopage
.end

**SMALL SIGNAL BIAS SOLUTION TEMPERATURE= 27.000 DEG C
```

Figure 3.3 (c). Output of SPICE simulator.

```
NODE VOLTAGE   NODE VOLTAGE   NODE VOLTAGE   NODE VOLTAGE

( 1)  10.0000 ( 2)  2.5000 ( 3)  0.0000 ( 4)  0.0000

      VOLTAGE SOURCE CURRENTS
      NAME           CURRENT
      vconst        -1.250E-01
      v2             6.250E-02
      v3             6.250E-02

      TOTAL POWER DISSIPATION  1.25E+00  WATTS
```

Figure 3.3 (c). Output of SPICE simulator (cont).

3.5 Source Output Variations

SPICE simulations can do a lot more than simple dc analysis of purely resistive circuits as will become evident in the later chapters. For the moment, we look at how the output of a source can be varied and the results presented in tabular and graphic forms.

The circuit shown in Figure 3.4 (a) has two voltage sources. The one on the right has a constant output of 5V. The current in R_3 due to this source is in a direction opposite from that due to the source on the left and hence can become zero. The output of the source on the left is varied until the current through the resistor R_3 becomes zero. For the current through R_3 to be zero, the voltage across it must also be zero by Ohm's law. Hence, we plot the voltage of the node 2 and check to see when it passes through the zero value.

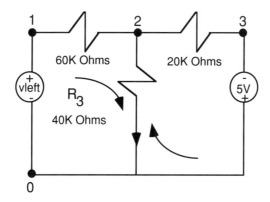

Figure 3.4 (a). Source output variations.

The circuit file is shown in Figure 3.4 (b). We have added a *control* line and two *output* lines to the file to attain our goal. The control line is

.dc vleft   14   16   0.2.

This line directs SPICE to vary the output of **vleft** from 14V to 16V in steps of 0.2V. The output lines are

.print dc v(2)

.plot dc v(2)

whose meanings should be obvious. The output is shown in Figure 3.4 (c), and we see that the voltage across **R**$_3$ becomes zero when **vleft** equals 15.1V approximately.

```
vs2r3   2voltage sources 3 resistors
*
*     consult Figure 3.4 (a)
*
vleft           1      0      dc     15
r1         1    2      60K
r2         2    3      20K
r3         2    0      40K
vright     0    3      dc     5
*
```

Figure 3.4 (b). SPICE circuit file.

```
* sweep output of vleft from 14V to 16V
* in increments of 0.2V
.dc vleft  14    16    0.2
*print values of voltage at node 2
*
.print dc v(2)
*
* plot voltage at node 2
*
.plot dc v(2)
.options nopage
.end
```

Figure 3.4 (b). SPICE circuit file (cont).

In Figure 3.4 (c) we observe that the **.print** command prints a table of values of the variables listed after the command. To obtain the value of the voltage drop across the 20K ohms we need a command such as **.print** v(2) v(2,3) or **.print** v(2,3).

Although numerical values are important, human comprehension is aided by graphs which show variations over a range at a glance. The **.plot** command is useful here. The graph created by the **.plot** command also prints the values of the first variable listed in the command, in this case v(2). It is obvious that **.plot** v(2) displays more information as an aid to human comprehension than does **.print** v(2). Hence, in our examples we have not made much use of the **.print** command and used **.plot** v(2) and **.plot** v(2,3) etc. instead.

```
vs2r3 2voltage sources 3 resistors

****   CIRCUIT DESCRIPTION
********************************************************

*
vleft              1      0      dc     15
r1          1      2      60K
r2          2      3      20K
r3          2      0      40K
vright      0      3      dc      5
```

Figure 3.4 (c). Output of SPICE simulation.

```
*
* sweep output of vleft from 14V to 16V
* in increments of 0.2V
*
.dc vleft  14    16     0.2
*
*print values of voltage at node 2
*
.print dc v(2)
*
* plot voltage at node 2
*
.plot dc v(2)
.options nopage
.end

*** DC TRANSFER CURVES              TEMPERATURE= 27.000 DEG C

vleft                V(2)
 1.400E+01           -1.818E-01
 1.420E+01           -1.455E-01
 1.440E+01           -1.091E-01
 1.460E+01           -7.273E-02
 1.480E+01           -3.636E-02
 1.500E+01           -6.653E-16
 1.520E+01            3.636E-02
 1.540E+01            7.273E-02
 1.560E+01            1.091E-01
 1.580E+01            1.455E-01
 1.600E+01            1.818E-01

*** DC TRANSFER CURVES              TEMPERATURE= 27.000 DEG C
```

Figure 3.4 (c). Output of SPICE simulation (cont).

```
vleft     V(2)
*)-----------    -2.000E-01 -1.000E-01 0.000E+00 1.000E-01 2.000E-01


1.400E+01  -1.818E-01  . *              .            .            .            .
1.420E+01  -1.455E-01  .       *        .            .            .            .
1.440E+01  -1.091E-01  .            *.               .            .            .
1.460E+01  -7.273E-02  .            .   *            .            .            .
1.480E+01  -3.636E-02  .            .        *  .                 .            .
1.500E+01  -6.653E-16  .            .             *              .            .
1.520E+01   3.636E-02  .            .             .  *           .            .
1.540E+01   7.273E-02  .            .             .       * .                 .
1.560E+01   1.091E-01  .            .             .            .*             .
1.580E+01   1.455E-01  .            .             .            .    *         .
1.600E+01   1.818E-01  .            .             .            .          *.


      JOB CONCLUDED
      TOTAL JOB TIME     2.41
1.480E+01  -3.636E-02  .            .        *  .                 .            .
1.500E+01  -6.653E-16  .            .             *              .            .
1.520E+01   3.636E-02  .            .             .  *           .            .
1.540E+01   7.273E-02  .            .             .       * .                 .
1.560E+01   1.091E-01  .            .             .            .*             .
1.580E+01   1.455E-01  .            .             .            .    *         .
1.600E+01   1.818E-01  .            .             .            .          *.


      JOB CONCLUDED
      TOTAL JOB TIME     2.41
```

Figure 3.4 (c). Output of SPICE simulation (cont).

3.6 Graphic Interface

The **.plot** command in SPICE uses a line printer with standard symbols or a console in the text mode to display a graph. Such graphs are not of the best visual quality and often not very easy to interpret. For this reason most SPICE packages provide a graphics postprocessor. Use of such a graphics processor produces graphs of better quality.

The names and calling conventions of such graphics processors vary from package to package. The reader should consult the SPICE system manual for these details. We are using a SPICE package provided by the **Microsim Corporation** of Laguna Hills, California. In this system, the graphics processor is called **Probe**™ and is activated by the **.probe** command in a circuit file. The reader will find many examples of this command used in the circuit files listed in this text.

3.7 Examples

Example 3.7.1

Consult the circuit diagram shown in Figure 3.5 (a). Using SPICE find the output of the source **ileft** which will cause i_3 to be zero.

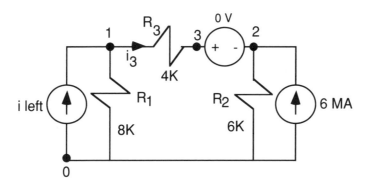

Figure 3.5 (a). SPICE example 1.

The circuit file as well as the output is shown in Figure 3.5 (b). Note that in order to obtain a graphical output we have used the graphic interface command for this particular version of SPICE called **.probe**. The control line causes the output of **ileft** to change from 0 mA to 8 mA in steps of 0.2 mA. The graphic output displays i_3 vs. the output of **ileft** and as can be seen, i_3 goes to zero when the output of **ileft** is approximately 4.5 mA.

```
2i3r.cir   2current sources & 3 resistors
*
*    consult Figure 3.5 (a)
*
ileft            0    1     dc    4mA
r1        0      1    8K
r3        1      3    4K
*
*   insert current meter vam in series with r3
*
vam       3      2    dc    0V
r2        2      0    6K
*
iright    0      2    dc    6mA
*
.dc  ileft       0    8E-3  2E-4
*
.probe   i(vam)
*
.options nopage
.end
```

Figure 3.5 (b). SPICE simulation output.

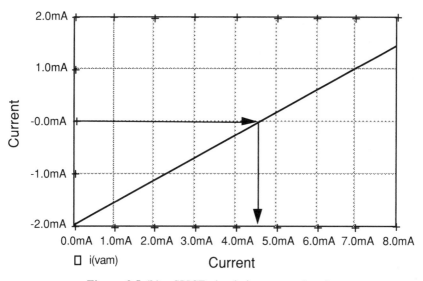

Figure 3.5 (b). SPICE simulation output (cont).

Example 3.7.2

Consult the circuit diagram shown in Figure 3.6 (a). Using SPICE simulation obtain a graphic plot of v_2 vs. i_S.

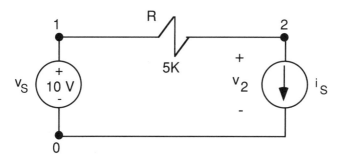

Figure 3.6 (a). SPICE example 2.

The circuit file as well as the output is shown in Figure 3.6 (b). When $i_S = 0$ according to Ohm's law the voltage drop in the 5 K resistor is zero. Hence, $v_2 = v_S = 10$ V. When $i_S = 2$ mA, the voltage drop in the 5 K resistor is 10 V and by Kirchhoff's voltage law $v_2 = 0$. These boundary conditions can be easily verified from the plot of v_2 vs. i_S in Figure 3.6 (b).

```
vir.cir
*
*      consult Figure 3.6 (a)
*
vleft              1       0       dc      10V
r1         1       2       5K
is         2       0       dc      1mA
*
.dc  is    0       2e-3      0.2e-3
*
.probe   v(2)
*
.options nopage
.end
```

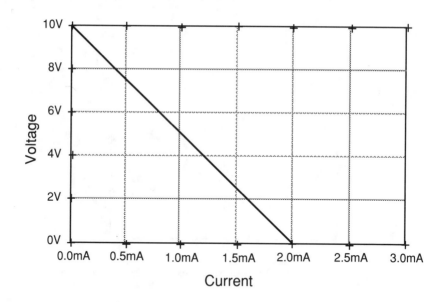

Figure 3.6 (b). SPICE simulation output.

3.8 Summary

- A circuit file contains description of a circuit, type of analysis to be performed, and the format of presentation of the output.

- Element lines are used to describe the positions and the values of elements in a circuit.

- Output of a dc source can be varied over any range of interest.

- **.print** creates a table of values.

- **.plot** creates a set of graphs. Graphs are easier to comprehend than tables.

- Graphic processor commands such as **.probe** can be used to obtain graphs of better visual quality.

3.9 Problems

3.1.   Write the SPICE circuit file for the purely resistive dc circuit shown. in Figure 3.7. Run it and verify the values computed by SPICE by separate manual computations.

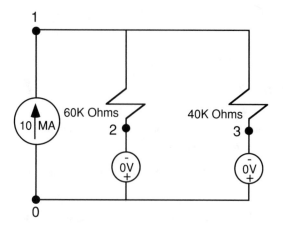

Figure 3.7. Current divider.

3.2.   Analyze the circuit shown in Figure 3.8 by SPICE simulation and manually
        verify the computed answers.

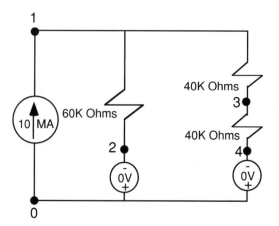

Figure 3.8.

3.3.   For the circuit shown in Figure 3.9, vary the output of the voltage source from
        7V to 12V in increments of 0.2V and plot the node voltage at node 2.

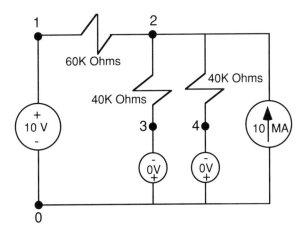

Figure 3.9.

3.4.   Using SPICE simulate the circuit shown in Figure 3.10, vary the output of v_S from 0 V to 5 V and plot i_S vs. v_S. By analyzing the circuit justify the values of i_S when v_S is 0 V and 5 V.

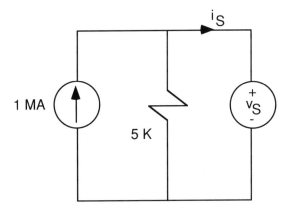

Figure 3.10.

3.5.   Simulate the circuit shown in Figure 2.16 for I_T = 8 mA, R_1 = R_2 = 6K ohms, and R_3 = 3K ohms. Compare the simulated value of v_3 with the value given by the expression obtained in Problem 2.1.

3.6.   Simulate the circuit shown in Figure 2.19 and obtain a value of **I** when **V**$_T$ = 10V, **R**$_1$ = 2K ohms, **R**$_2$ = 3K ohms, **R**$_3$ = 4K ohms and **R** = 1K ohms. Compare this value of **I** with the value given by the expression derived in Problem 2.4.

3.7.   In the simulated circuit of Problem 3.6, attach an independent voltage source across the terminals marked a and b (see Figure 2.19) Vary the output of this voltage source and obtain a graph of its terminal current vs. terminal voltage.

3.8.   In the simulated circuit of Problem 3.7, attach an independent current source across the terminals marked a and b (see Figure 2.19) Vary the output of this current source and obtain a graph of its terminal voltage vs. terminal current.

3.9.   Simulate the circuit shown in Figure 2.18 and attach an independent voltage source across the terminals marked a and b. Assume **R**$_1$ = **R**$_2$ = 2K ohms, **R**$_3$ = 6K ohms and **R**$_4$ = 4K ohms. Vary the output of this voltage source and obtain a graph of its terminal current vs. terminal voltage. Compute the ratio V/I from this graph and compare with the result of Problem 2.3.

3.10.   Simulate the circuit shown in Figure 2.18 and attach an independent current source across the terminals marked a and b. Assume **R**$_1$ = **R**$_2$ = 2K ohms, **R**$_3$ = 6K ohms and **R**$_4$ = 4K ohms. Vary the output of this current source and obtain a graph of its terminal voltage vs. terminal current. Compute the ratio V/I from this graph and compare with the result of Problem 2.3

4

CONTROLLED SOURCES

Controlled sources are important circuit elements for most electronic circuits. The output of such a source is either a voltage or a current which is controlled by voltages and/or currents at other parts of a circuit. Active circuits incorporating transistors are modelled by using controlled sources. In SPICE, other frequently used circuit elements such as variable resistors can also be modelled by controlled sources.

4.1 Controlled Sources

Let v_c or i_c denote the output voltage or current of a controlled source. Then in general, v_c or i_c are given by equations of the form:

$$v_c = f(v_1, v_2, \ldots, i_1, i_2, \ldots),$$
$$i_c = g(v_1, v_2, \ldots, i_1, i_2, \ldots).$$

Here the *control* variables $v_1, v_2, \ldots, i_1, i_2, \ldots$, etc. are voltages and currents at other parts of a circuit and f and g are nonlinear functions. Controlled sources are called linear if f and g are *linear* functions of their arguments and, furthermore, useful special cases arise when there is only one control variable controlling the output of a source.

4.1.1 Linear Controlled Sources

Since the output as well as the control variable can be of two types, there are four different types of controlled sources. These are

1. Voltage controlled voltage source (vcvs): $v_c = \mathbf{k}\, v_j$

   $\mathbf{k}$ is the voltage gain.

2.   Current controlled current source (cccs): $i_c = k\ i_j$

k is the current gain.

3.   Voltage controlled current source (vccs): $i_c = k\ v_j$

k is the transconductance.

4.   Current controlled voltage source (ccvs): $v_c = k\ i_j$

k is the transresistance.

The graphical symbols of these sources are shown in Figure 4.1 The element lines in a SPICE circuit file for each of these sources are given in the next section.

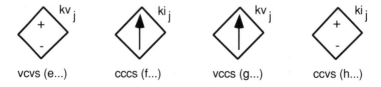

Figure 4.1. Controlled sources.

4.1.2   SPICE Element Lines

The name of a vcvs starts with the letter **e**. Those of cccs, vccs, and ccvs start with the letters **f**, **g**, and **h** respectively. The element lines for a vcvs and a vccs are of the form:

ename	n+	n-	nc+	nc-	k
gname	n-	n+	nc+	nc-	k

n+ and n- are the nodes between which the controlled source is connected. nc+ and nc- are the nodes defining the location of the control variable (voltage) in the circuit. The last parameter on the line gives the value of **k** the voltage gain or the transconductance as the case may be.

The element lines for a cccs and a ccvs have slightly different formats than the ones shown above. These are current controlled sources and hence, the control variable in each case is a current in some branch of a circuit. As discussed in Chapter 3, to measure the current in a branch in a SPICE simulation of a circuit we use a dead voltage source. Now we assign a name to such a dead voltage source and use this name in the element definition line of a cccs or a ccvs. Each dead source has also to be defined on a separate element line of its own. The element lines look as follows:

fname	n-	n+	deadsource	k
deadsource	nc+	nc-	dc	0V
hname	n+	n-	deadsource	k
deadsource	nc+	nc-	dc	0V

4.2 Examples

4.2.1 VCCS

Figure 4.2 gives the diagram of a circuit with a voltage controlled current source. The current output of the controlled source is given by the expression $k\ v_2$ where the controlling voltage v_2 is the voltage drop across R_2. Note that v_2 is an unknown quantity. The problem is to find an expression for i_4, the current through the resistor R_4.

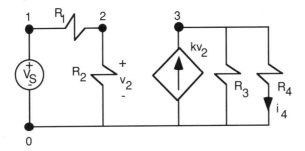

Figure 4.2. Example of a VCCS.

By current division

$$i_4 = k \, v_2 \, R_3 \, / \, (R_3 + R_4). \tag{4.2.1}$$

By voltage division

$$v_2 = V_S \, R_2 \, / \, (R_1 + R_2). \tag{4.2.2}$$

Substituting v_2 from (4.2.2) in (4.2.1) we get,

$$i_4 = k \, V_S \, R_2 \, R_3 \, / \, ((R_3 + R_4) \, (R_1 + R_2)).$$

4.2.2   CCVS

Figure 4.3 shows a circuit with a current controlled voltage source. In order to find an expression for v_4 we proceed as follows:

By current division

$$i_2 = I_S \, R_1 \, / \, (R_1 + R_2). \tag{4.2.3}$$

By voltage division and substitution of i_2 from (4.2.3) we have,

$$v_4 = k \, i_2 \, R_4 \, / \, (R_3 + R_4)$$
$$= k \, I_S \, R_1 \, R_4 \, / \, ((R_1 + R_2) \, (R_3 + R_4)). \tag{4.2.4}$$

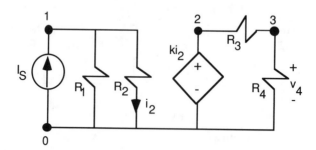

Figure 4.3. Example of a CCVS.

Figure 4.4 shows another circuit containing a current controlled voltage source. An expression for i_2 can be derived in the following manner:

Using Kirchhoff's voltage law around the loop excluding the current source we have

$$i_1 (k - R_1) + i_2 R_2 = 0. \tag{4.2.5}$$

Using Kirchhoff's current law at node 1 we get,

$$i_1 + i_2 = I_S. \tag{4.2.6}$$

Solving for i_2 from (4.2.5) and (4.2.6) we derive,

$$i_2 = I_S (R_1 - k) / (R_1 + R_2 - k). \tag{4.2.7}$$

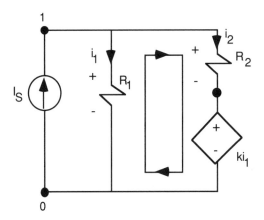

Figure 4.4.   Second CCVS example.

4.2.3   VCVS

Figure 4.5 shows a voltage divider with a voltage controlled voltage source in place of a load resistor. Derivation of an expression for i_0 is shown below.

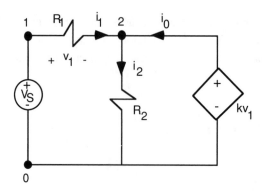

Figure 4.5. Example of a VCVS.

Using Kirchhoff's voltage law around the loop containing the two voltage sources we have $\mathbf{k}\ \mathbf{v}_1 + \mathbf{v}_1 = \mathbf{V}_S$. This equation gives us an expression for $\mathbf{v}_1$. By Ohm's law $\mathbf{i}_1 = \mathbf{v}_1 / \mathbf{R}_1$ and $\mathbf{i}_2 = \mathbf{k}\ \mathbf{v}_1 / \mathbf{R}_2$. Using Kirchhoff's current law at node 2 we have $\mathbf{i}_0 = \mathbf{i}_2 - \mathbf{i}_1$. Finally substituting for $\mathbf{i}_2$, $\mathbf{i}_1$ and $\mathbf{v}_1$ in sequence in the above equation for $\mathbf{i}_0$ we have

$$\mathbf{i}_0 = \mathbf{V}_S\ (\mathbf{k}\ \mathbf{R}_1 - \mathbf{R}_2) / ((\mathbf{k} + 1)\ \mathbf{R}_1\ \mathbf{R}_2). \tag{4.2.8}$$

4.3 Validation of Expressions

First let us consider the circuit in Figure 4.4 using a ccvs. If the parameter $\mathbf{k}$ is set to zero, then the output of the ccvs becomes zero and we have a simple current divider. In this case

$$\mathbf{i}_2 = \mathbf{I}_S\ \mathbf{R}_1 / (\mathbf{R}_1 + \mathbf{R}_2).$$

If $\mathbf{k} = \mathbf{R}_1$, then the voltage drop across $\mathbf{R}_2$ becomes zero. Hence, by Ohm's law $\mathbf{i}_2 = 0$. These two cases can be verified from the expression of $\mathbf{i}_2$ in (4.2.7). As an exercise the reader should determine what happens to $\mathbf{i}_2$ as $\mathbf{k}$ becomes very large.

Next let us consider the circuit in Figure 4.5. Here if $k = 0$ the voltage drop across R_2 becomes zero and by Ohm's law $i_2 = 0$. By Kirchhoff's current law at node 2, $i_0 =$ -i_1. Since the output of the vcvs is zero, by Kirchhoff's voltage law, $v_1 = V_S$. Hence i_0 = -V_S / R_1. This case can be easily derived from the expression of i_0 in (4.2.8).

4.4 Examples Using SPICE

Figure 4.6 shows the circuit file as well as the graph of i_4 vs. V_S of the circuit shown in Figure 4.2.. This circuit uses a vccs. Substituting the parameter values given in the circuit file in the expression for i_4 obtained in Example 4.2.1 we have $i_4 = 1.9\ V_S$. The value of V_S ranges from 0 to 5 V and the graph shows the corresponding values of i_4.

Figure 4.7 shows the circuit file as well as the output voltage v_4 of the circuit shown in Figure 4.3. This circuit uses a ccvs. To monitor the controlling current i_2 we have added a dead source in series with R_2. The output of the independent current source varies from -1 mA to +1 mA and the corresponding values of v_4 are shown in a graph.

```
vccs.cir   how to describe a vccs in spice
*
*      consult Figure 4.2.A
*
vs             1      0      dc     8V
r1             1      2      3K
r2             2      0      6K
* a vccs between nodes 3 and 0
gs             0      3      2      0      4
r3             3      0      5K
r4             3      4      2K
* vd in series with r4 to measure i4
vd             4      0      dc     0V
.dc  vs    0    5    0.2
.probe   i(vd)
.options nopage
.end
*
```

Figure 4.6. VCCS example output.

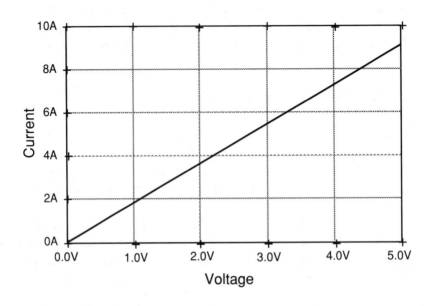

Figure 4.6. VCCS example output (cont).

```
ccvs.cir   circuit with a ccvs
*
*     consult Figure 4.3
*
is          0      1      0A
r1          1      0      4K
r2          1      4      4K
* dead source for monitoring i2 in r2
vd          4      0      dc      0V
* ccvs between nodes 0 and 2 with k = 10
hs          2      0      vd      10
r3          2      3      4K
r4          3      0      4K
```

Figure 4.7. CCVS example output.

```
* vary is output from -1 mA to +1 mA
.dc    is  -1E-3   1E-3   1E-4
.probe v(3,0)
.options nopage
.end
*
```

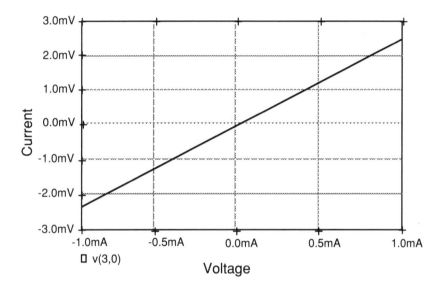

Figure 4.7. CCVS example output (cont).

4.5 Nonlinear Controlled Sources

The functions f and g appearing in the controlling equations of *nonlinear* controlled sources are in general nonlinear. In SPICE, such nonlinear functions are approximated by polynomials. The key word used is **poly**(n) where n denotes the number of variables in the polynomial. If **x** and **y** are two variables used in **poly**, then **poly**(2) is

$$\textbf{poly}(2) = \; p_0 +$$
$$p_1 \, \textbf{x} + p_2 \, \textbf{y} +$$
$$p_3 \, \textbf{x}^2 + p_4 \, \textbf{x} \, \textbf{y} + p_5 \, \textbf{y}^2 +$$
$$p_6 \, \textbf{x}^3 \, \textbf{y} + p_7 \, \textbf{x}^2 \, \textbf{y} + ... \hspace{2cm} (4.5.1)$$

Nonlinear controlled sources are valuable tools of circuit simulation in SPICE as shown in the following section.

4.6   Variable Resistor

Numerical analysis of a circuit for a single set of parameter values does not provide enough information about the circuit behavior in general. Typically we wish to know how a circuit will behave when some parameter value is varied over a given range. For purely resistive circuits, this means we need the capability of defining variable resistors and of controlling variations in their values in an automatic manner. Of course, variable resistors will be useful elements in other circuits as well. In SPICE, variable resistors can be created by using nonlinear controlled sources.

4.6.1   CCVS as a Resistor

The output equation of a ccvs is given by $v_c = \textbf{k} \, \textbf{i}_j$. If the controlling current $\textbf{i}_j$ happens to be the current $\textbf{i}_c$ through the ccvs itself, then across its terminals the ccvs behaves as a constant resistor $\textbf{R} = v_c / \textbf{i}_c = \textbf{k}$. The element lines needed in a circuit file to create such a resistor (between nodes n+ and n-) out of a ccvs are given below.

hresistor	n+	m	deadsource	k
deadsource	m	n-	dc	0V

Figure 4.8 shows the arrangement of the ccvs and the dead source in a graphical manner.

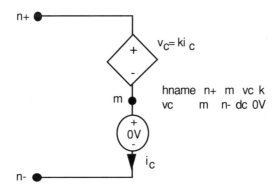

Figure 4.8. CCVS as a resistor.

4.6.2   Nonlinear CCVS

Let us now consider a nonlinear ccvs whose output v_C is controlled by its own terminal current i_C and another current i_S through a dead source v_S. We define **poly** in (4.5.1) by setting $p_0 = p_1 = p_2 = p_3 = 0$ and $p_4 = \mathbf{k}$. Then

$$v_C = \mathbf{poly}(2)$$
$$= \mathbf{k}\, i_C\, i_S,$$

and $\mathbf{R} = v_C / i_C = \mathbf{k}\, i_S$. The value of $\mathbf{R}$ depends on i_S and hence, $\mathbf{R}$ is a variable resistor whose value can be varied by sweeping the output of an independent current source i_S over any suitable range. The element lines needed in the circuit file to define a variable resistor (between nodes n+ and n-) are shown below.

hvar	n+	m	poly(2)	deadsource	vs	0 0 0 0 k
deadsource	m	n-	dc	0V		
vs	a	b	dc	0V		
is	b	a	dc	1A		

Figure 4.9 shows the arrangement of the ccvs and the i_S source in a graphical manner.

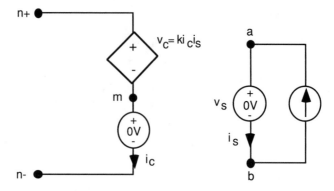

Figure 4.9. Nonlinear CCVS as a variable resistor.

Nonlinear controlled sources can also be used to create linear controlled sources with variable **k**. Such elements are useful in circuit analysis. As an example we can write v_c = **k** i_j i_s as v_c = (**k** i_s) i_j where **k** is becomes a variable transresistance whose value can be changed by changing the output of the source i_s.

4.7 Circuit Simulation with Variable Resistors

4.7.1 Example: Power Transfer

In Chapter 2 we analyzed a circuit with a variable resistor for maximizing power (consult Example 2.9.2). The circuit diagram is given in Figure 2.9. Let us analyze this circuit again to demonstrate that the power of SPICE simulation often goes beyond what can be achieved by manual analysis.

In Example 2.9.2 we found that the power in the resistor **R** attains a maximum when **R** = **R**$_S$. We can easily verify this known fact by using SPICE. However, SPICE can be used to plot power in **R** as a function of **R**. Such a graph tells us at a glance how *sensitive* the power value is to variations in **R** near its peak.

The Figure 4.10 shows the circuit file and the graph of power in **R** vs. values of **R**. The element line **hvar** describes the variable resistor whose value is controlled by the current through the dead source **vc** produced by the independent source **ic**. The value of **k** in the **hvar** line is 1 K, or 1000. The output of **ic** in the .**dc** analysis line goes from 0 to 20 in steps of 0.5. Hence, the value of **hvar** changes from 0 K to 20 K in steps of 0.5 K. If we mentally change the label **A** along the independent axis of the graph to **K**, then the values of **R** can be read directly from the graph.

To plot the graph of power in **R** we stored the values of the current in **R**, i.e., **i(vd)** and the voltage across **R**, i.e., **v(1,2)** defined in the circuit file. The **Probe**™ graphics processor allows us to plot arithmetic expressions of circuit variables defined in the circuit file. The plot of power is obtained by plotting **i(vd) * v(1,2)**.

From the graph shown in Figure 4.10 we see that maximum power is attained when **R** = 5 K since the value of **R**$_S$ is set to 5 K in the circuit file. However, we also see how the power varies with variations in **R** about the peak value. The peak value of power is 31 mW approx. If the value of **R** is increased 100% to 10 K the power value is approximately 28 mW, a decrease of 9.7%. However, only a 50% decrease in the value of **R** causes the same percentage decrease in power. Hence, if we are not certain about the value of **R**$_S$ we should overestimate it to obtain close to maximum power in **R**. An underestimation can cause serious loss of power instead. Such useful information is hard to obtain by visual inspection of the expression of power.

```
power.cir power in a variable resistor
*
*     consult Figure 2.9
*
is          0       1       5mA
rs          1       0       5K
* create variable resistor between nodes 0 & 1
hvar                1       2       poly(2)     vd    vc 0 0 0 0 1K
vd          2       0       dc          0V
vc          3       0       dc          0V
ic          0       3       dc          1A
* vary output of ic to vary value of hvar
```

Figure 4.10. Power in a resistor.

```
.dc     ic   0    20   0.5
* store i and v values to plot power
.probe i(vd) v(1,2)
.end
*
```

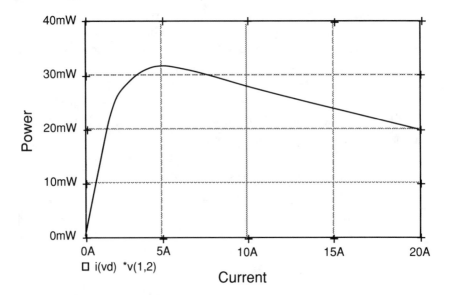

Figure 4.10. Power in a resistor (cont).

4.7.2 Example: Analysis of Expressions

In Section 2.10 of Chapter 2 we have an analysis of an expression obtained in Example 2.9.7. There we tested for the correctness of an expression for i_3,

$$i_3 = (I_1 R_1 - I_2 R_2) / (R_1 + R_2 + R_3),$$

by varying the value of R_2 from zero to infinity. We can test our analysis by using SPICE to simulate this situation. Figure 4.11 shows the circuit file (consult Figure 2.14) and the graph of i_3 vs. R_2. Using the expression derived for i_3 when R_2 is zero in Section 2.10 and the values of the circuit parameters given in Figure 4.11 we have $i_3 = 2$ mA. Similarly when R_2 is infinity, $i_3 = -4$ mA. From the graph in Figure 4.11 we see that $i_3 = 2$ mA when R_2 is zero and i_3 tends to -4 mA as R_2 tends to infinity.

4.7.3 Example: VCVS with Variable Gain

Figure 4.12 shows the circuit file for the circuit shown in Figure 4.5 and analyzed in Section 4.3. Using a nonlinear vcvs we have created a linear vcvs with variable k. The control line causes the value of k to be increased starting at 0. As the graph shows the value of i_0 starts at $-V_S / R_1 = -2.5$ mA and goes through 0 when $k = R_2 / R_1 = 2$ and approaches $V_S / R_2 = 1.25$ mA with increasing k. The current i_1 starts at V_S / R_1 and approaches 0 as k increases since an increase in k v_1 causes v_1 to decrease by Kirchhoff's voltage law.

```
Fig214.cir   check analysis of
* Example 2.9.7 in Chapter 2
*
*    consult Figure 2.14
*
i1          0      1      4mA
r1          1      0      4K
r3          1      3      4K
* vd1 measures current i3 in r3
vd1         3      2      dc     0V
* r2 is a variable resistor between nodes 2 & 0
hr2         2      4      poly(2)  vd2  vs 0 0 0 0 1K
vd2         4      0      dc     0V
vs          5      0      dc     0V
is          0      5      dc     1mA
```

Figure 4.11. Circuit file and graph of i_3 vs. $\mathbf{R}_2$.

```
* current source on the right
i2         0     2     4mA
* vary is output to get variable r2
.dc    is   0    60    0.5
* store values of i3 for plot
.probe i(vd1)
.options nopage
.end
*
```

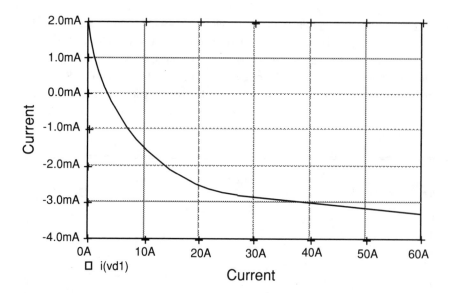

Figure 4.11. Circuit file and graph of i_3 vs. $\mathbf{R}_2$ (cont).

```
vcvs.cir   vcvs with variable gain k
*
*     consult Figure 4.5
*
vs          1    0    dc    10V
r1          1    3    4K
* dead source for measuring i1
vd          3    2    dc    0V
r2          2    0    8K
es          2    0    poly(2)  1  2  4  0  0 0 0 0 1
* output of vc controls variable k
vc          4    0    dc    0V
* complete loop across vc by r0
r0          4    0    1
* vary output of vc to vary k
.dc  vd  0  8   0.2
.probe i(es) i(vd)
.options nopage
.end
*
```

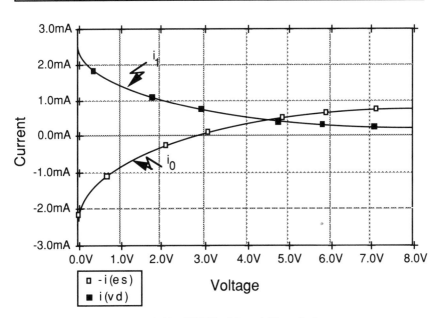

Figure 4.12. VCVS with variable gain k.

4.8 Summary

- Controlled sources can be linear or nonlinear. Linear controlled sources are used in the analysis of electronic circuits. These are vcvs, cccs, vccs, and ccvs.

- Names of linear controlled sources in SPICE start with the letters **e, f, g**, or **h**. See Section 4.1.2 for details of these element lines.

- The nonlinear control function of a nonlinear controlled source is approximated by a polynomial in SPICE (see Section 4.5).

- Variable resistors can be simulated in SPICE by using a nonlinear ccvs (see Section 4.6).

4.9 Problems

4.1. Find an expression for the current i_1 in the circuit shown in Figure 4.13. Join nodes a and b by a zero resistance wire and repeat previous analysis.

4.2. Find an expression for **i** in the circuit shown in Figure 4.14.

4.3. Find an expression for v_2 in the circuit shown in Figure 4.15.

4.4. For the circuit shown in Figure 2.21, let $V_T = 5V$, $R_1 = R_2 = 3K$ ohms, $R_5 = 1$ ohm and $R = 5K$ ohms. Execute a SPICE simulation of the circuit for values of R_3 in the range 4K to 6K ohms in steps of 0.2K ohms and obtain a graph of **I** vs. R_3. Test the expression for balance condition against the simulation results.

4.5. Execute a SPICE simulation of the circuit shown in Figure 2.22 and obtain a graph of v_R vs. **R**. Test the expression for the maximum and the minimum values of v_R obtained in Problem 2.7 against the simulation results. For this simulation assume that $V_T = 5$ V, $R_1 = 7K$ ohms and $R_2 = 4K$ ohms.

4.6.   In the circuit shown in Figure 2.23, assume that I_T = 7 mA, R_1 = 2K ohms, and R_2 = 3K ohms. Execute a SPICE simulation of the circuit and obtain a graph of i_R vs. **R**. Check the expression for the maximum value of i_R against the simulation results.

4.7.   Rerun the SPICE simulation of Problem 4.3 and obtain a graph of power in **R** vs. **R**. At what value of **R** does the power in **R** attain its maximum?

4.8.   In a simple voltage divider circuit with resistors R_1 and R_2, the output of the voltage source is 12V. The output voltage across R_2 (with no load resistor in parallel) is required to be 4V. The power values in R_1 and R_2 are not to exceed 0.288 watts. Find the minimum values of R_1 and R_2 necessary to implement this voltage divider.

4.9.   Assume that a load resistor **R** is connected in parallel with R_2 of the voltage divider of Problem 4.5. Find an expression for the power in **R**. From this expression find the value of **R** by manual analysis that corresponds to the maximum value of the power in **R**. Verify your results by SPICE simulation.

4.10.   In a simple voltage divider circuit with resistors R_1 and R_2, the output of the voltage source is 12V. R_1 = 4K ohms and a load resistor **R** = 6K ohms is connected in parallel with R_2. The power in **R** is 0.006 watts. Find the value of R_2.

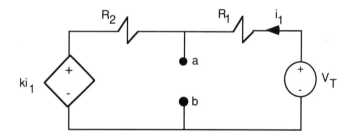

Figure 4.13.

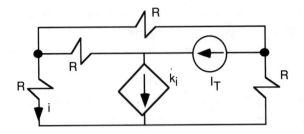

Figure 4.14.

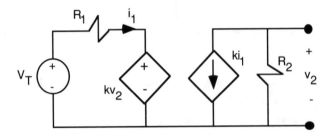

Figure 4.15.

5

LINEAR CIRCUIT THEOREMS

A general circuit element can be thought of as a black box with a number of terminals. Some of the voltages and currents applied to these terminals are called input signals. Other terminal voltages and terminal currents are called output signals. A circuit element is linear between its input and output terminals if the output signals are linear functions of the input signals. A circuit constructed out of linear circuit elements behaves as a linear circuit. Several useful theorems can be derived for linear circuits which are presented in this chapter.

5.1 Linear Circuits

A circuit as a black box is shown in Figure 5.1. The input signals are labelled x, y, and z and the output signal is w. In general w is a nonlinear function of x, y, and z as shown below:

$$w = f(x, y, z). \tag{5.1.1}$$

A circuit is linear if and only if the function f is linear in its arguments, i.e., we can write

$$w = a\,x + b\,y + c\,z, \tag{5.1.2}$$

where a, b, and c are constants. A resistor is a linear circuit element if its terminal voltage/current relationship is given by Ohm's law since this law specifies a linear relation. A circuit constructed of linear resistors behaves as a linear circuit. Other linear circuit elements will be introduced in later chapters.

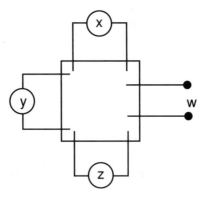

Figure 5.1. A circuit as a black box.

5.2 Superposition

A powerful method of analysis of linear circuits with multiple independent sources is based on the *principle of superposition*. This principle follows naturally from the linearity of a circuit. As an illustration, let us use the circuit shown in Figure 5.1 where x, y, and z are assumed to be independent input sources. Let w_x denote the output signal when the only input source is x. Also w_y and w_z are defined in a similar manner. Then using the linear input/output relation (5.1.2) we have $w_x = a\, x$, $w_y = b\, y$, and $w_z = c\, z$. Hence, we can write

$$w = w_x + w_y + w_z. \tag{5.2.1}$$

Each partial output signal w_x, w_y, or w_z is obtained with only one independent input source in circuit. Hence, the output signal of a linear circuit with multiple independent sources can be obtained by superposing the partial output signals contributed by each independent source acting alone.

Since the principle of superposition requires us to analyze a circuit with only one independent source at a time, we must have some means of deactivating the other independent sources in the circuit. Deactivating a source means to set its terminal output to zero. For a voltage source this can be done by joining its output terminals together through an element of zero resistance. This is called *short-circuiting* a voltage source. A current source can be deactivated by disconnecting one or both of its terminals from a circuit. This is called *open-circuiting* a current source. Consult Figure 5.2 for a graphical illustration of these operations.

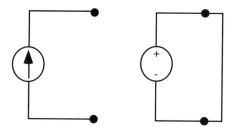

Figure 5.2. Open and short circuiting sources.

5.3 Multisource Circuits

A circuit with multiple independent sources is shown in Figure 5.3 (a). This circuit is to be analyzed to find an expression for the current i_2.

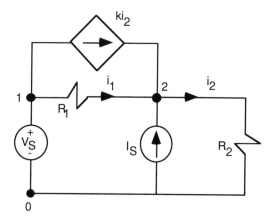

Figure 5.3 (a). Multisource circuit.

First we short-circuit the independent voltage source. The transformed circuit is shown in Figure 5.3 (b). Using current division we can write

$$i_{21} = (I_S + k\ i_{21})\ R_1\ /\ (R_1 + R_2).$$ (5.3.1)

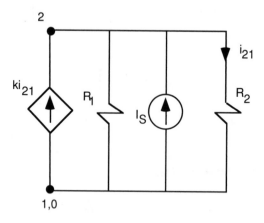

Figure **5.3 (b).**   Transformed multisource circuit.

From (5.3.1) using simple algebra we obtain

$$i_{21} = I_S\ R_1\ /\ (R_2 + (1 - k)\ R_1).$$ (5.3.2)

Next we open-circuit the independent current source. The transformed circuit is shown in Figure 5.3 (c).

Using Kirchhoff's current law at node 2 we find that the current through R_1 is $(1 - k)$ i_{22}. Using Kirchhoff's voltage law in the loop containing V_S, R_1, and R_2 we can write

$$i_{22}\ (R_2 + (1 - k)\ R_1) + V_S = 0.$$ (5.3.3)

Again using simple algebra we obtain from (5.3.3)

$$i_{22} = -\ V_S\ /\ (R_2 + (1 - k)\ R_1).$$ (5.3.4)

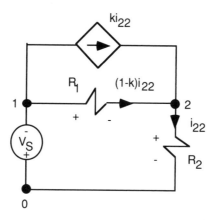

Figure 5.3 (c). Open-circuit with independent current source.

Finally, superposing (5.3.2) and (5.3.4) we have

$$\mathbf{i}_2 = \mathbf{i}_{21} + \mathbf{i}_{22},$$
$$= (\mathbf{I_S} \, \mathbf{R}_1 - \mathbf{V_S}) \, / \, (\mathbf{R}_2 + (1 - \mathbf{k}) \, \mathbf{R}_1). \qquad (5.3.5)$$

Note that only independent sources are deactivated; controlled sources are left in the circuit as they are.

5.4 Open and Short Circuits

The next section developes two fundamental theorems of linear circuit analysis. In this connection it becomes necessary to consider open and short circuits. We think of a circuit as if it is contained in a box. Its only connection to the outside world is through certain terminals which are brought out of the box for this purpose. Consult Figure 5.4 for a graphic demonstration. If no circuit element is connected between a pair of such terminals in the outside world, then we call the pair to be an *open* circuit. If a pair of such terminals are joined together through an element of zero resistance, then we call the pair to be a *short* circiut.

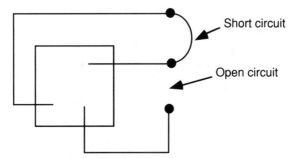

Figure 5.4. Open and short circuit.

The analytic concepts of open and short circuits should not be confused with their graphic symbols. For an open (short) circuit to exist, the terminal current (voltage) must be zero regardless of the graphic symbol in use at a particular moment.

5.5 Equivalent Circuits

Consider the two circuits enclosed in boxes shown in Figure 5.5. The terminal voltage/current relations for the two circuits are

$$v_1 = f(i_1),$$
$$v_2 = g(i_2).$$

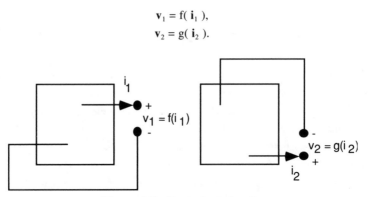

Figure 5.5. Equivalent circuits.

We call the two circuits *equivalent* if f and g are identical functions. From the outside world, two circuits are equivalent only with respect to a fixed set of terminals and only if the terminal voltage/current relations are the same for both circuits.

5.6 Norton's and Thevenin's Theorems

An equivalent circuit can be a very useful tool in circuit analysis if its internal structure is much simpler than that of the circuit to which it is equivalent. Thevenin's and Norton's theorems show us how to obtain extremely simple equivalent circuits of linear circuits of arbitrary complexity. Consider a linear circuit shown enclosed in a box in Figure 5.6 whose terminal voltage/current relation by assumption is linear.

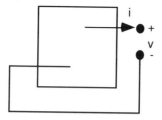

Figure 5.6.

From the outside world we can connect an appropriate independent voltage (current) source to these terminals without changing the terminal current (voltage). The transformed circuits are shown in Figure 5.7.

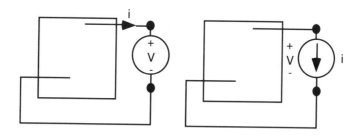

Figure 5.7.

Since the circuit is linear, we can find its terminal current (voltage) by the principle of superposition discussed earlier. First by short (open) circuiting the outside voltage (current) source we find the terminal current (voltage) to be i_{sc} (v_{oc}). Consult Figure 5.8.

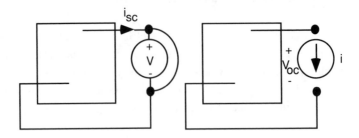

Figure 5.8.

Next we assume that all the internal independent sources are deactivated. Consult Figure 5.9.

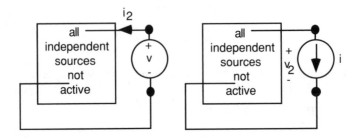

Figure 5.9.

The terminal current (voltage) i_2 (v_2) is given by v / R_{eq} ($i R_{eq}$) where R_{eq} is the equivalent internal resistance of the circuit with all internal sources deactivated. Using superposition we get $i = i_{sc} - i_2$ ($v = v_{oc} - v_2$) which can be rewritten as

$$\text{Norton's equation:} \quad i = i_{sc} - v / R_{eq}, \tag{5.6.1}$$
$$\text{Thevenin's equation:} \quad v = v_{oc} - i R_{eq}. \tag{5.6.2}$$

Norton's and Thevenin's equations describe the terminal voltage/current relation of the same circuit. Hence, the **i** in Norton's equation must be the same as the **i** in Thevenin's equation. Substituting the expression for **i** from Norton's equation into Thevenin's equation and rearranging terms we have

$$\mathbf{R}_{eq} = \mathbf{v}_{oc} / \mathbf{i}_{sc}. \tag{5.6.3}$$

Any circuit whose terminal voltage/current relation is given by either Norton's equation or Thevenin's equation is equivalent to the circuit of Figure 5.6. Fortunately, there are extremely simple circuits satisfying these criteria and these are called the Norton and the Thevenin equivalent circuits respectively. These equivalent circuits are shown in Figure 5.10.

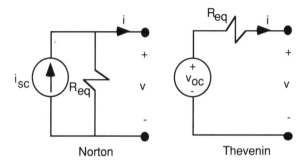

Figure 5.10.

Graphical representations of Norton's and Thevenin's equations are frequently helpful for comprehension and are shown in Figure 5.11.

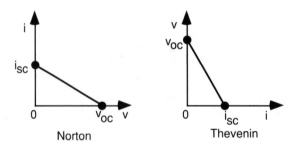

Figure 5.11.

5.7 Examples

Example 5.7.1

As our first example let us consider the circuit shown in Figure 5.12. To find its equivalent circuits we have to obtain the expressions for v_{OC} and i_{SC} for the pair of terminals marked a and b.

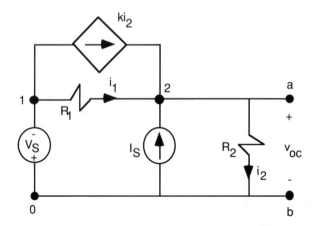

Figure 5.12.

Clearly $v_{OC} = i_2 R_2$ and we have already obtained an expression for i_2 by superposition (5.3.5). Hence,

$$v_{OC} = (I_S \, R_1 - V_S) \, R_2 / (R_2 + (1 - k) \, R_1). \qquad (5.7.1)$$

To find i_{sc} we short the terminals a and b as shown in Figure 5.13.

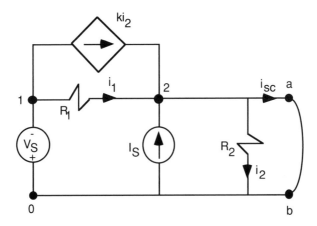

Figure 5.13.

Since the voltage drop across R_2 is zero, $i_2 = 0$. The controlled current source becomes an open circuit. By inspection

$$i_{sc} = I_S - (V_S / R_1). \qquad (5.7.2)$$

Hence, from (5.7.1) and (5.7.2) we derive

$$\begin{aligned} R_{eq} &= v_{OC} / i_{sc} \\ &= R_1 \, R_2 / (R_2 + (1 - k) \, R_1). \end{aligned} \qquad (5.7.3)$$

Either of the circuits shown in Figure 5.6.E can be used as equivalent circuits of the circuit shown in Figure 5.12 with the expressions for v_{OC}, i_{sc}, and R_{eq} given above.

Example 5.7.2

Consider the Wheatstone bridge circuit shown in Figure 5.14 and suppose we want to find an expression for the current across the bridge when it is unbalanced.

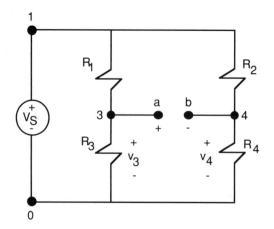

Figure 5.14.

One approach is to replace the bridge by its Thevenin equivalent circuit between terminals a and b. By voltage division and Kirchhoff's law

$$\mathbf{v}_{oc} = \mathbf{V}_S \, \mathbf{R}_3 \, / \, (\mathbf{R}_1 + \mathbf{R}_3) \; - \; \mathbf{V}_S \, \mathbf{R}_4 \, / \, (\mathbf{R}_2 + \mathbf{R}_4),$$
$$= \mathbf{V}_S \, (\mathbf{R}_2 \, \mathbf{R}_3 - \mathbf{R}_1 \, \mathbf{R}_4) \, / \, ((\mathbf{R}_1 + \mathbf{R}_3) \; (\mathbf{R}_2 + \mathbf{R}_4)). \qquad (5.7.4)$$

Note that for a balanced bridge $\mathbf{v}_{oc} = 0$. To find $\mathbf{R}_{eq}$ we use an alternate approach. $\mathbf{R}_{eq}$ is the equivalent resistance of the circuit between terminals a and b when all the internal sources have been deactivated. Short circuiting the voltage source in the Wheatstone bridge we obtain the circuit shown in Figure 5.15.

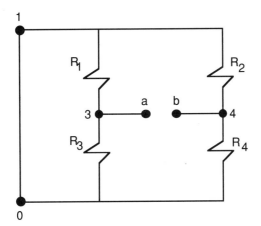

Figure 5.15.

The resistance between terminals a and b is by inspection

$$\mathbf{R_{eq}} = (\mathbf{R}_1\,\mathbf{R}_3\,/\,(\mathbf{R}_1 + \mathbf{R}_3)) + (\mathbf{R}_2\,\mathbf{R}_4\,/\,(\mathbf{R}_2 + \mathbf{R}_4)). \qquad (5.7.5)$$

Either the Thevenin or the Norton equivalent circuit can now be easily obtained. It is trivial to obtain an expression for the current across the unbalanced bridge from these equivalent circuits.

Example 5.7.3

In this example we show a third alternate method for deriving $\mathbf{R_{eq}}$. This method is particularly useful when the circuit in question has internal controlled sources. As an example consider the circuit shown in Figure 5.12. If following the approach in the second example we deactivate all the internal independent sources we obtain the circuit shown in Figure 5.16.

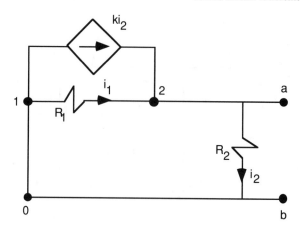

Figure 5.16.

Since there are no independent sources to drive the circuit, the current i_2 is obviously zero. The controlled source output becomes zero and the source behaves as an open circuit. The equivalent resistance between terminals a and b becomes $R_1 R_2 / (R_1 + R_2)$ which from Example 1 we know to be in error.

The problem here is due to the fact that deactivating all internal independent sources also deactivates any internal controlled source that may be present in a circuit. To keep the internal controlled sources operational we introduce a test source across terminals a and b. Consult Figure 5.17.

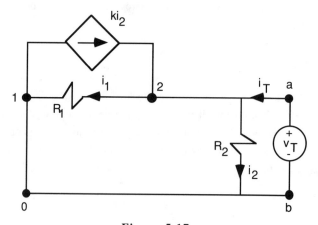

Figure 5.17.

In general this test source can be either a voltage source or a current source. To simplify analysis we have used a voltage source in Figure 5.17. Between a and b by inspection $\mathbf{R}_{eq} = v_T / i_T$.

By inspection of Figure 5.17 we obtain by Ohm's law $i_1 = v_T / \mathbf{R}_1$, $i_2 = v_T / \mathbf{R}_2$ and by Kirchhoff's law $i_T = v_T / \mathbf{R}_1 + v_T / \mathbf{R}_2 - k\, v_T / \mathbf{R}_2$. The ratio $\mathbf{R}_{eq} = v_T / i_T$ can be easily obtained from this expression and is the same as that obtained earlier in Example 5.7.1.

5.8 Validation of Expressions

Norton and Thevenin equivalent circuits are simpler than the original circuits. Hence, they should provide greater insights into the terminal behaviors of circuits. The expressions for v_{oc} and i_{sc} depend on the circuit parameters. We can use these expressions to analyze how the terminal behavior of a circuit changes with changes in circuit parameter values.

5.8.1 Wheatstone Bridge Sensitivity

Let us consider the Wheatstone bridge discussed in Example 5.7.2. We assume that $\mathbf{R}_4$ is an unknown resistance whose value is to be determined by balancing the bridge and that the value of $\mathbf{R}_3$ is varied until balance condition is reached. The resistance of the current measuring ammeter (consult Figure 2.7) is assumed to be very small. Then the current through this ammeter is essentially the short circuit current. Using the expressions obtained for v_{oc} (5.7.4) and $\mathbf{R}_{eq}$ (5.7.5) we obtain the following expression for i_{sc}.

$$i_{sc} = V_S\,(\mathbf{R}_2\,\mathbf{R}_3 - \mathbf{R}_1\,\mathbf{R}_4) / (\mathbf{R}_1\,\mathbf{R}_3\,(\mathbf{R}_2 + \mathbf{R}_4) + \mathbf{R}_2\mathbf{R}_4\,(\mathbf{R}_1 + \mathbf{R}_3)). \qquad (5.7.6)$$

Under condition of balance $i_{sc} = 0$ and $\mathbf{R}_4 = (\mathbf{R}_2 / \mathbf{R}_1)\,\mathbf{R}_3$. Let us use ß to denote the ratio $(\mathbf{R}_2 / \mathbf{R}_1)$. Then i_{sc} becomes

$$i_{sc} = V_S(ß\mathbf{R}_3 - \mathbf{R}_4) / (\mathbf{R}_2\,(\mathbf{R}_3 + \mathbf{R}_4) + (1 + ß)\mathbf{R}_3\,\mathbf{R}_4). \qquad (5.7.7)$$

When $R_3 = 0$, $i_{sc} = -V_S / R_2$ which is independent of ß. As R_3 approaches infinity i_{sc} approaches ß $V_S / (R_2 + (1 + ß)R_4)$. If this value of i_{sc} is close to zero, then it will be difficult to detect the zero crossing of i_{sc} at balance condition while varying R_3. However, this value depends on ß and hence it can be controlled by a proper choice of ß. Manual analysis at this point becomes rather complex and so we use SPICE to simulate the circuit.

Figure 5.18 shows the circuit file for the Wheatstone bridge with a variable resistor for R_3. The value for ß in this file is set to 10. Two other runs were made with ß set to 1 and 0.1 respectively. The graphs of i_{sc} show that higher values of ß make it easier to detect the zero crossing of i_{sc} and provide a more sensitive bridge. This is another example of how SPICE simulation can be used to obtain insight into the behavior of circuit variables under varying conditions.

```
wheat.cir wheatstone bridge circuit
*
*     consult Figure 2.7 or 5.14
*
vin         1     0     dc      10V
r1          1     2     0.1K
r2          1     3     1K
* value of beta = r2/r1 = 10
r4          3     0     4K
* source between nodes 2 & 3 for balance
* detection by current measurement
vd1         2     3     dc      0V
* variable resistor r3
hr3         2     5     poly(2)    vd2    vs 0 0 0 0 1K
vd2
vs
ls
*
.dc         is    0     1       0.1
.probe      i(vd1)
.end
*
```

Figure 5.18. Circuit file for Wheatstone bridge with variable resistor.

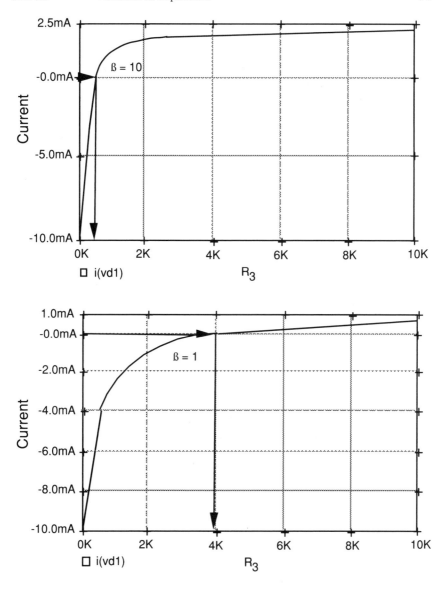

Figure 5.18. Circuit file for Wheatstone bridge with variable resistor (cont).

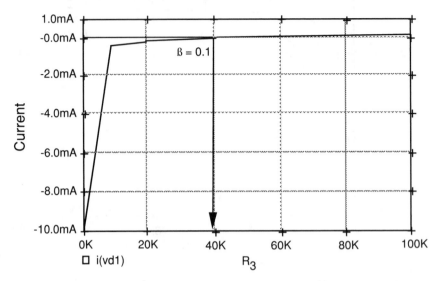

Figure 5.18. Circuit file for Wheatstone bridge with variable resistor (cont).

5.8.2   Example 5.7.1

Figure 5.12 shows the circuit analyzed in Example 5.7.1. Let us test the plausibility of the expressions we obtained in that case.

The circuit becomes simple if the controlled source is deactivated. So we set $k = 0$. The short circuit current is independent of the controlled source and so it remains

$$i_{sc} = I_S - (V_S / R_1).$$

Applying the method presented in Example 5.7.2 to this simple circuit we obtain by inspection

$$R_{eq} = R_1 R_2 / (R_1 + R_2).$$

Since $v_{oc} = i_{sc} R_{eq}$ we have

$$v_{oc} = (I_S R_1 - V_S) R_2 / (R_1 + R_2).$$

The same expression results after substitution of $k = 0$ in the original expression of v_{oc} obtained earlier (5.7.1).

5.9 Equivalent Circuits via. SPICE Simulations

The expressions for v_{oc}, i_{sc}, and R_{eq} of any Norton or Thevenin equivalent circuit depend on the parameters of the original linear circuit. Hence, at least for simple circuits which are amenable to manual analysis, we can obtain some insights into the circuit's terminal behavior as parameter values are varied. It is difficult to analyze complex circuits in this way; however, SPICE simulation can be used to obtain these equivalent circuits provided numerical analysis is acceptable for our purpose.

To find the Norton equivalent circuit across any pair of terminals we need to find i_{sc} and R_{eq}. Using SPICE simulation we can obtain a graphical representation of Norton's equation as shown in Figure 5.11. We connect an independent voltage source across the terminals as shown in Figure 5.7, vary its output and plot the terminal current. The reader should be able to figure out a similar method for obtaining a graphical representation of Thevenin's equation.

5.9.1 Example 5.7.1

Consider the circuit shown in Figure 5.12 and analyzed in Example 5.7.1. Figure 5.19 shows the SPICE circuit file for this circuit as well as the graph of its Norton's equation. From Figure 5.19 we observe that $V_S = 1$ V, $I_S = 2$ mA, $R_1 = 1K$ ohms, $R_2 = 2K$ ohms, and $k = 2$. Substituting these values in the expressions (5.7.1) and (5.7.2) we have $v_{oc} = 2$ V and $i_{sc} = 1$ mA. The graph of Norton's equation shown in Figure 5.19 gives us exactly these same values and by direct computation $R_{eq} = 2$ V $/ 1$ mA $= 2K$ ohms.

```
Fig 512.cir finding norton equivalent
*
*      consult Figure 5.12
*
vs              0     1       dc      1V
r1              1     2       1K
r2              2     3       2K
vd              3     0       dc      0V
is              0     2       dc      2mA
* ccvs between nodes 1 & 2
fs              1     2       vd      2
* connect vtest between nodes 2 & 0
vtest                2     0       dc      1V
* vary output of vtest to get
*norton's v-i equation
.dc  vtest  0  2  0.1
.probe i(vtest)
.options nopage
.end
*
```

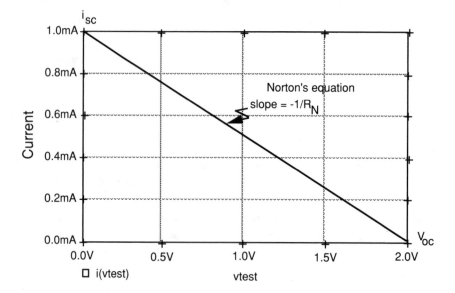

Figure 5.19. SPICE output of Norton's equation.

5.9.2 Wheatstone Bridge

Figure 5.20 shows the SPICE circuit file for a Wheatstone bridge as well as the graph of its Thevenin's equation. Since we are after Thevenin's equation, the independent source across the terminals is a current source. Using the parameter values given in the circuit file in (5.7.4) and (5.7.5) the reader should verify the correctness of the graph of Thevenin's equation shown in Figure 5.20.

```
wheat.cir   wheatstone bridge
* finding thevenin equivalent
* of a whatstone bridge
*
*     consult Figure 5.14
*
vs          1     0      dc      10V
r1          1     3      0.1K
r3          3     0      0.05K
r2          1     4      1K
r4          4     0      4K
* put current source between nodes 3 & 4
itest             4     3      dc      1mA
* vary output of itest
.dc   itest   0   5.6e-3   0.1e-3
.probe v(4,3)
.options nopage
.end
*
*
```

Figure 5.20.   SPICE output of Thevenin's equation.

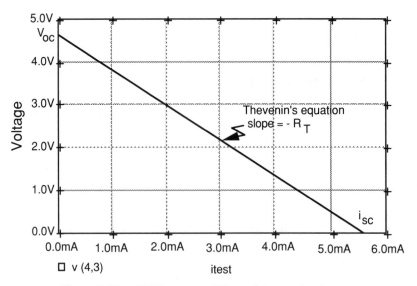

Figure 5.20. SPICE output of Thevenin's equation (cont).

5.10 Source Transformations

Source transformations are special cases of Norton's and Thevenin's theorems. However, they are often very useful in circuit analysis. According to source transformation rules any part of a circuit with the structure shown in Figure 5.21 part (a) can be replaced by its equivalent circuit shown in part (b) and *vice versa*.

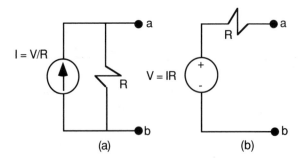

Figure 5.21. Source transformations.

5.11 Summary

•   *Principle of superposition*: The total output signal is the sum of the output signals each produced by an independent source acting alone. This principle applies to all linear circuits.

•   Two circuits are *equivalent* with respect to a given set of terminals if their terminal voltage/current relations are identical.

•   Thevenin's and Norton's theorems provide the means for constructing simple equivalent circuits of arbitrary linear circuits.

5.12 Problems

5.1.   Consider a simple voltage divider with source voltage V_T, and series resistors R_1 and R_2. Remove R_2 from the circuit and replace the remaining circuit by its Norton equivalent circuit (Note: this is an instance of using source transformation). Reconnect R_2 to the Norton equivalent and find expressions for the current through and the voltage across R_2. Compare the last expression with (2.6.3a). Find an expression for the current through R_1 in the equivalent circuit. Is this expression correct (for the original voltage divider)? If not, then give a simple explanation of why not.

5.2.   Using superposition find an expression for the current through the controlled source in the circuit shown in Figure 5.22.

5.3.   Using superposition find an expression for the voltage across the controlled source in the circuit shown in Figure 5.23.

5.4.   Using superposition find an expression for the current through the independent voltage source in the circuit shown in Figure 5.24.

5.5.   The circuit shown in Figure 5.25 has a voltage of 10V across the terminals a and
        b. When a 10K ohms resistor is connected across a and b, the current in the
        resistor is 0.5 mA. Does this indicate a violation of Ohm's law? If not, then
        explain why not. What would be the current in a 5K ohms resistor if it is
        connected across the terminals a and b?

5.6.   Consider another circuit that also looks like the circuit of Figure 5.25. A 3K
        ohms and an 8K ohms resistor connected across the terminals a and b have
        currents 4 mA and 2 mA in them respectively. What would be the current in a
        6K ohm resistor connected across terminals a and b?

5.7.   (Maximum power theorem) A variable resistor R is connected across the
        terminals a and b of a circuit that looks like the circuit shown in Figure 5.25.
        Find an expression for the value of R at which the power in R attains its
        maximum value.

5.8.   For the circuit shown in Figure 5.22, assume that $V_T = 5V$, $I_T = 5$ mA, $k = 0.5$,
        $R_1 = 2K$ ohms, and $R_2 = 4K$ ohms. Using SPICE simulation, find the Norton
        equivalent of this circuit across the terminals marked a and b.

5.9.   For the circuit shown in Figure 5.24, assume that $V_T = 5V$, $I_T = 5$ mA, $k =$
        0.0001, and $R = 2K$ ohms. Using SPICE simulation, find the Thevenin
        equivalent of this circuit across the terminals marked a and b.

5.10.   The terminal voltage/current relationship of a linear circuit is shown in Figure
        5.26. Another circuit, which draws 2A at a terminal voltage of 5V is to be
        connected to the first circuit. How many ways can this be done while satisfying
        the terminal voltage and current requirements of the second circuit? Which way
        uses the least amount of power, and how much power does it use?

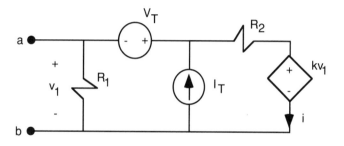

Figure 5.22.

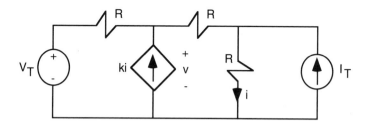

Figure 5.23.

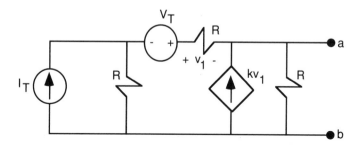

Figure 5.24.

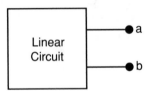

Figure   5.25.

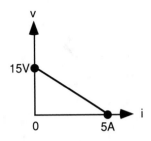

Figure 5.26.

6

CIRCUIT EQUATIONS

In the previous chapters we have analyzed circuits by the intelligent application of Kirchhoff's and Ohm's laws. Many derived results such as the voltage and the current division formulas or Thevenin's and Norton's theorems are also very useful on their own. However, this heuristic approach becomes harder to maintain as the circuits become more complex.

In this chapter we derive two methods of circuit analysis that are entirely mechanical in their operations. These are the *node equation* and the *mesh equation* methods. Both these methods set up sets of simultaneous equations in the circuit variables. The node equation approach is based on Kirchhoff's current law. The mesh equation approach makes use of Kirchhoff's voltage law. So either one of these two laws is sufficient for the complete analysis of a circuit. Usually a circuit has much fewer essential nodes than meshes and hence, the node equation approach is used more frequently. Also, in the context of a computer program, it is easier to describe a circuit in terms of its nodes and circuit elements than in terms of its meshes. Nodes are the primary entities for describing circuits in a SPICE simulation model. We have stressed the use of the node equations as the primary approach in this chapter.

6.1 Nodes and Meshes

A node in a circuit is a junction point where circuit elements are connected together. Figures 6.1 (a) and 6.1 (b) show two circuits with their nodes numerically labelled. A node where more than two elements are connected together is called an *essential node*. In Figure 6.1 (a) nodes 0 and 1 are essential nodes but nodes 2 and 3 are not. All three nodes in Figure 6.1 (b) are essential nodes.

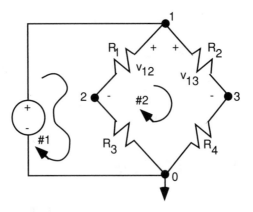

Figure 6.1 (a). Circuit with two essential nodes.

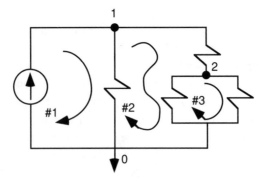

Figure 6.1 (b). Circuit with all nodes essential.

A node in a circuit is called a *reference node* if the voltages at all the other nodes are measured with respect to the voltage at that node. By definition the voltage at the reference node is zero. Any arbitrary node in a circuit can be selected as its reference node. However, some choices of reference nodes reduce the subsequent effort needed to analyze the circuit. These choices will be discussed in the following sections. In Figures 6.1 (a) and 6.1 (b), the nodes labelled 0 are taken to be the reference nodes. The arrow head shown in each figure at node 0 is often used to indicate a reference node.

The voltage at a node of a circuit, called a *node voltage*, is always measured with respect to the voltage at the reference node of the circuit. A node voltage is a different concept from a terminal voltage of a circuit element. In Figure 6.1 (a), the node voltage v_1 at node 1 is not the terminal voltage of either resistor R_1 or resistor R_2. These terminal voltages are $v_{12} = v_1 - v_2$ and $v_{13} = v_1 - v_3$, where v_2 and v_3 are the node voltages at nodes 2 and 3 respectively. In general since a circuit element is connected between two nodes, its terminal voltage is the difference between those two node voltages. If one of these two nodes happen to be the referrence node, then the terminal voltage equals the node voltage of the other node. In Figure 6.1 (a), v_2 and v_3 are the terminal voltages of the resistors R_3 and R_4 respectively.

Figure 6.2 shows a resistor R connected between two nodes 1 and 2. The direction of the current in R can be assumed to be either from node 1 to node 2 (i_{12}) or from node 2 to node 1 (i_{21}). In the first case we write $i_{12} = (v_1 - v_2) / R$ and in the second case $i_{21} = (v_2 - v_1) / R$. Note that as expected from the Figure 6.2 $i_{12} = - i_{21}$.

A closed path or loop in a circuit is traced out by starting at any node and then moving through a set of connected circuit elements in such a manner that we return to the starting node without repeating any intermediate node. A *mesh* is a loop containing no other loops in it. In Figure 6.1 (a) there are two meshes, (0,1,2,0) and (0,2,3,0). There are three meshes in Figure 6.1 (b), (0,1,0), (0,1,2,0), and (0,2,0).

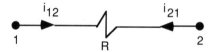

Figure 6.2. Resistor connecting two nodes.

Each mesh is assumed to have a single current flowing around it which is called a *mesh current*. A mesh current is in general different from the terminal current of a circuit element which is included in the mesh. In general a circuit element is part of more than one mesh. Hence, its terminal current is the algebraic sum of all the mesh currents of all the meshes that share this circuit element. If a circuit element is in only one mesh, then its terminal current equals that mesh current.

6.2 Node Equations

The values of all the circuit variables can be obtained if all the node voltages in the circuit are known. All circuit elements are connected between at least two nodes. Hence, their terminal voltages are given by the difference of those two node voltages. The terminal currents can then be computed from the terminal voltage/current relations of the circuit elements.

To find all the node voltages we need to write some equations in them and solve these equations simultaneously for the node voltage values. The necessary equations can be written by applying Kirchhoff's current law at each node. Thus Kirchhoff's current law is all that we need to analyze a circuit if we use the node equation approach. The process is entirely mechanical and can be programmed for computers. The easiest way to describe the construction of the node equations is by considering specific types of circuits. This is done in the following sections.

6.3 Resistive Circuits with Voltage Sources

Figure 6.3 shows a circuit made of resistors and multiple voltage sources. To write the node equations for this circuit we need to label its nodes as shown in Figure 6.3. Any node in this circuit can be chosen as the reference node. However, to simplify the resulting node equations, it is wise to choose a node to which multiple voltage sources are connected. In Figure 6.3, the node 0 is chosen as the reference node.

The purpose of writing node equations is to find the node voltages. If one terminal of a circuit element is connected to the reference node, then the node voltage at the other terminal equals the terminal voltage of the circuit element. If the circuit element happens to be a voltage source, then this node voltage is known immediately. In Figure 6.3, $v_1 = V_1$ and $v_4 = -V_2$.

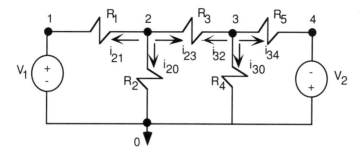

Figure 6.3. Circuit composed of resistors and multiple voltage sources.

Next we apply Kirchhoff's current law at nodes 2 and 3. Note that these are the only essential nodes of the circuit other than the reference node itself. The essential nodes of a circuit are the only nodes whose node equations are needed. To apply Kirchhoff's current law, let us arbitrarily but consistently assume that *all currents leaving a node are positive*. At node 2 we can write

$$\mathbf{i}_{21} + \mathbf{i}_{20} + \mathbf{i}_{23} = 0,$$

or $$(\mathbf{v}_2 - \mathbf{v}_1) / \mathbf{R}_1 + (\mathbf{v}_2 - \mathbf{v}_0) / \mathbf{R}_2 + (\mathbf{v}_2 - \mathbf{v}_3) / \mathbf{R}_3 = 0,$$

or $$\mathbf{v}_2 (1/\mathbf{R}_1 + 1/\mathbf{R}_2 + 1/\mathbf{R}_3) - \mathbf{v}_3 (1/\mathbf{R}_3) = \mathbf{V}_1 / \mathbf{R}_1. \qquad (6.3.1)$$

Similarly at node 3 we can write

$$\mathbf{i}_{32} + \mathbf{i}_{30} + \mathbf{i}_{34} = 0,$$

or $$(\mathbf{v}_3 - \mathbf{v}_2) / \mathbf{R}_3 + (\mathbf{v}_3 - \mathbf{v}_0) / \mathbf{R}_4 + (\mathbf{v}_3 - \mathbf{v}_4) / \mathbf{R}_5 = 0,$$

or $$- \mathbf{v}_2 (1/\mathbf{R}_3) + \mathbf{v}_3 (1/\mathbf{R}_3 + 1/\mathbf{R}_4 + 1/\mathbf{R}_5) = - \mathbf{V}_2 / \mathbf{R}_5 \qquad (6.3.2)$$

The two node equations are (6.3.1) and (6.3.2) and these are to be solved simultaneously for the unknown node voltages $\mathbf{v}_2$ and $\mathbf{v}_3$. Figure 6.4 shows another resistive circuit with two voltage sources. However, in this circuit the two sources do not share a common node. So we arbitrarily pick node 0 to be the reference node. As a result we have $\mathbf{v}_1 = \mathbf{V}_1$. At the essential node 2, using Kirchhoff's current law we get

$$\mathbf{i}_{21} + \mathbf{i}_{20} + \mathbf{i}_{23} = 0,$$

or $$(\mathbf{v}_2 - \mathbf{v}_1) / \mathbf{R}_1 + (\mathbf{v}_2 - \mathbf{v}_0) / \mathbf{R}_2 + \mathbf{i}_{23} = 0. \qquad (6.3.3)$$

Here we realize that i_{23} is an unknown current through the source V_2 and hence must somehow be eliminated from (6.3.3). Although node 3 is not an essential node we write its node equation since i_{23} goes into node 3.

$$i_{32} + i_{30} = 0,$$

or
$$i_{32} + (v_3 - v_0) / R_3 = 0. \tag{6.3.4}$$

Note that $i_{23} = - i_{32}$. Therefore by adding (6.3.3) and (6.3.4) we have

$$(v_2 - v_1) / R_1 + (v_2 - v_0) / R_2 + (v_3 - v_0) / R_3 = 0,$$

or
$$v_2 (1/ R_1 + 1/ R_2) + v_3 (1/ R_3) = V_1 / R_1. \tag{6.3.5}$$

Now we have only one equation (6.3.5) but two unknowns v_2 and v_3. However, from Figure 6.4 we see that v_2 and v_3 are related by $v_3 = v_2 - V_2$. Substituting this expression for v_3 in (6.3.5) we obtain a single equation in one unknown v_2, the voltage of the essential node 2. Thus first v_2 and then v_3 can be determined in sequence.

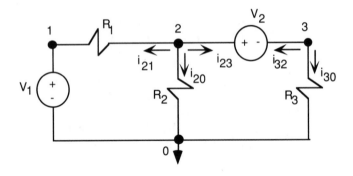

Figure 6.4. Resistive circuit with two voltage sources.

6.4 Resistive Circuits with Current Sources

A current source in a circuit directly specifies the value of a current at two nodes to which it is connected. Since node equations are written by applying Kirchhoff's current law at nodes, the presence of a current source at a node obviously simplifies that node equation. The best choice of a reference node in this case is any node with a large number of resistors connented to it. Figure 6.5 shows a circuit with two current sources. The node equation at node 1 is

$$- \mathbf{I}_1 + \mathbf{i}_{12} + \mathbf{i}_{10} = 0,$$

or
$$- \mathbf{I}_1 + (\mathbf{v}_1 - \mathbf{v}_2) / \mathbf{R}_3 + (\mathbf{v}_1 - \mathbf{v}_0) / \mathbf{R}_1 = 0,$$

or
$$\mathbf{v}_1 (1/\mathbf{R}_1 + 1/\mathbf{R}_3) - \mathbf{v}_2 (1/\mathbf{R}_3) = \mathbf{I}_1. \qquad (6.4.1)$$

Similarly the node equation at node 2 is

$$\mathbf{I}_2 + \mathbf{i}_{21} + \mathbf{i}_{20} = 0,$$

or
$$\mathbf{I}_2 + (\mathbf{v}_2 - \mathbf{v}_1) / \mathbf{R}_3 + (\mathbf{v}_2 - \mathbf{v}_0) / \mathbf{R}_2 = 0,$$

or
$$- \mathbf{v}_1 (1/\mathbf{R}_3) + \mathbf{v}_2 (1/\mathbf{R}_2 + 1/\mathbf{R}_3) = - \mathbf{I}_2. \qquad (6.4.2)$$

The two node equations are (6.4.1) and (6.4.2) and these can be solved simultaneously for the unknown node voltages $\mathbf{v}_1$ and $\mathbf{v}_2$.

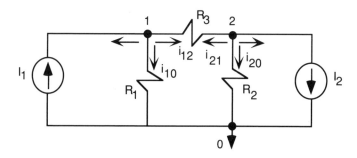

Figure 6.5. Resistive circuit with two current sources.

6.5 Resistive Circuits with Controlled Sources

Although there are four types of controlled sources we shall consider only a CCVS. The manner of handling other controlled sources should be clear from this discussion as well as the previous sections. Figure 6.6 shows a circuit with a voltage source, a current source, and a CCVS. Node 0 is the reference node and $v_1 = - V_1$. The node equations are

At node 2: $v_2 (1/R_1 + 1/R_2 + 1/R_3) - v_3 (1/R_3) = - V_1 / R_1.$ (6.5.1)

At node 3: $- v_2 (1/R_3) + v_3 (1/R_3 + 1/R_4) + i_{34} = 0.$ (6.5.2)

At node 4: $i_{43} + v_4 (1/R_5) = - I_1.$ (6.5.3)

i_{34} in (6.5.2) and i_{43} in (6.5.3) are unknown currents but $i_{34} = - i_{43}$. Hence adding (6.5.2) and (6.5.3) together we have

$$- v_2 (1/R_3) + v_3 (1/R_3 + 1/R_4) + v_4 (1/R_5) = - I_1.$$ (6.5.4)

Now we have two equations (6.5.1) and (6.5.4) in three unknowns v_2, v_3, and v_4. However, from Figure 6.6 we see that $v_4 = v_3 + k\ i_{32} = v_3 + k\ (v_3 - v_2) / R_3 = v_3 (1 + k / R_3) - v_2 (k / R_3)$. Substituting this expression for v_4 in (6.5.4) we obtain two equations which can be solved for v_2 and v_3.

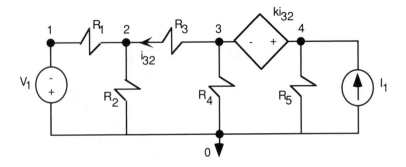

Figure 6.6. Resisitive circuit with a voltage source, a current source and a CCVS.

6.6 Mesh Equations

To write the mesh equations of a circuit we first identify the meshes and assign a mesh current to each mesh. Figure 6.7 shows a circuit with three mesh currents i_1, i_2, and i_3. In general a circuit element is part of more than one mesh. Hence, its terminal current is the algebraic sum of all the mesh currents of all the meshes that share this circuit element. If a circuit element is in only one mesh, then its terminal current equals that mesh current.To obtain the mesh equations we apply Kirchhoff's voltage law to each mesh in turn. For the first mesh we have

$$i_1 \, \mathbf{R}_1 + (i_1 - i_2) \, \mathbf{R}_2 - \mathbf{V}_1 = 0,$$

which can be written as

$$i_1 \, (\mathbf{R}_1 + \mathbf{R}_2) - i_2 \, \mathbf{R}_2 = \mathbf{V}_1. \tag{6.6.1}$$

The equation for the second mesh is

$$\mathbf{k} \, i_{20} + (i_2 - i_3) \, \mathbf{R}_3 + (i_2 - i_1) \, \mathbf{R}_2 = 0. \tag{6.6.2}$$

Note that $i_{20} = i_1 - i_2$. Substituting for i_{20} in (6.6.2) and rearranging terms we get

$$i_1 \, (\mathbf{k} - \mathbf{R}_2) + i_2 \, (\mathbf{R}_2 + \mathbf{R}_3 - \mathbf{k}) - i_3 \, \mathbf{R}_3 = 0. \tag{6.6.3}$$

By inspection of the third mesh we have

$$i_3 = -\mathbf{I}_1. \tag{6.6.4}$$

Hence, the third mesh current i_3 is given directly by (6.6.4). Substituting for i_3 in (6.6.3) and solving simultaneously with (6.6.1) we can obtain the other two mesh currents i_2 and i_1. Once all the mesh currents are known, the branch currents and the branch voltages can be easily obtained.

We note that since Kirchhoff's voltage law is used, the outputs of voltage sources are simply added to the other voltage terms in a mesh equation. In case of a controlled source, further manipulation of its output expression may become necessary to eliminate extra unknowns. The mesh current of a mesh containing a current source not shared with any other mesh is given directly as the output of that current source. If the current source is shared with other meshes, then the algebraic sum of the mesh currents equal the output of the current source.

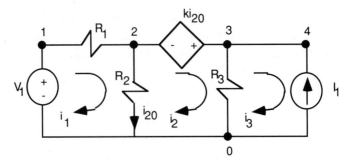

Figure 6.7. Circuit with three mesh currents.

6.7 Examples

Example 6.7.1

The circuit is shown in Figure 6.8. In order to derive its node equations we label the nodes as shown in Figure 6.8 and take node 0 to be the reference node. It follows that $v_1 = V_1$ and the node equation at node 2 is

$$(v_2 - v_1) / R_1 + (v_2 - v_3) / R_3 - I_1 + k\, v_{12} = 0. \qquad (6.7.1)$$

From the circuit diagram $v_{12} = v_1 - v_2$. Substituting this expression for v_{12} in (6.7.1) and rearranging terms we get

$$v_2\,(1/R_1 + 1/R_3 - k) - v_3\,(1/R_3) = I_1 + V_1(1/R_1 - k). \qquad (6.7.2)$$

The node equation for node 3 is

$$v_3\,(1/R_2) + (v_3 - v_2) / R_3 - k\, v_{12} = 0, \qquad (6.7.3)$$

which after substitution of $v_{12} = v_1 - v_2$ and rearrangement of terms becomes

$$- v_2\,(1/R_3 - k) + v_3\,(1/R_2 + 1/R_3) = k\, V_1. \qquad (6.7.4)$$

(6.7.2) and (6.7.4) are the node equations for this circuit.

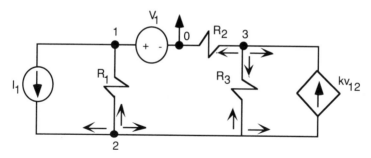

Figure 6.8.

Example 6.7.2

For the bridge circuit shown in Figure 6.9 we label the nodes as shown in the Figure and assume node 0 to be the reference node. Immediately we have $v_1 = V_1$. The node equation at node 2 is

$$v_2 (1/R_1 + 1/R_3 - 1/R_5) - v_3 (1/R_5) = V_1 (1/R_1), \tag{6.7.5}$$

and the node eqation at node 3 is (by symmetry)

$$- v_2 (1/R_5) + v_3 (1/R_2 + 1/R_4 + 1/R_5) = V_1 (1/R_2). \tag{6.7.6}$$

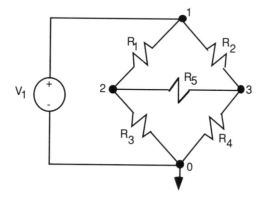

Figure 6.9. Bridge circuit.

Example 6.7.3

To write the mesh equations for the circuit shown in Figure 6.10 we label the nodes, the meshes, and the mesh currents as shown. The mesh equation for the first mesh is

$$i_1 (R_1 + R_2) - i_2 R_2 = V_1. \qquad (6.7.7)$$

In the second mesh we note that $i_2 = - k\, i_{30} = - k\, (i_2 - i_3)$ which can be rewritten as

$$i_2 (1 + k) - k\, i_3 = 0. \qquad (6.7.8)$$

Finally the third mesh gives us

$$i_3 = - I_1. \qquad (6.7.9)$$

Note that the presence of a CCCS tend to simplify mesh equations.

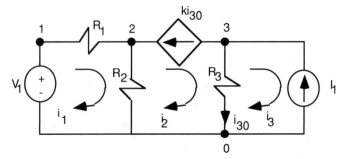

Figure 6.10.

Example 6.7.4

The three mesh currents for the bridge circuit are shown in Figure 6.11. The mesh equations are

$$i_1 (R_1 + R_3) - i_2 R_1 - i_3 R_3 = V_1, \qquad (6.7.10)$$

and

$$- i_1 R_1 + i_2 (R_1 + R_2 + R_5) - i_3 R_5 = 0, \qquad (6.7.11)$$

$$- i_1 R_3 - i_2 R_5 + i_3 (R_3 + R_4 + R_5) = 0. \qquad (6.7.12)$$

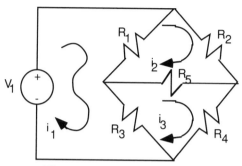

Figure 6.11. Bridge circuit.

6.8 Summary

•   A reference node in a circuit is any node whose voltage is assumed to be zero. The voltages of all the other nodes are measured with respect to the reference node.

•   An essential node of a circuit is any node to which more than two circuit elements are connected.

•   A node voltage is the voltage of a node with respect to the voltage of the reference node. A node voltage is the terminal voltage of a circuit element if one terminal of the element is connected to the reference node.

•   Node equations are obtained by applying Kirchhoff's current law to each essential node of a circuit.

•   A mesh is a loop containing no other loops. A mesh current is a current assumed to flow around a mesh. The terminal current of a circuit element shared by several meshes is the algebraic sum of all those mesh currents. Mesh equations are obtained by applying Kirchhoff's voltage law to each mesh in a circuit.

6.9 Problems

6.1.   For the ten circuits shown below in Figure 6.12, write their node equations.

6.2.   For the ten circuits shown below in Figure 6.12, write their mesh equations.

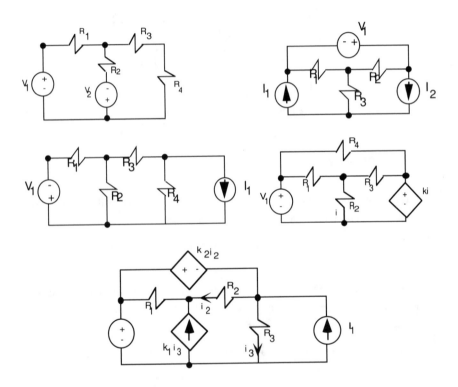

Figure 6.12. Circuits for problems.

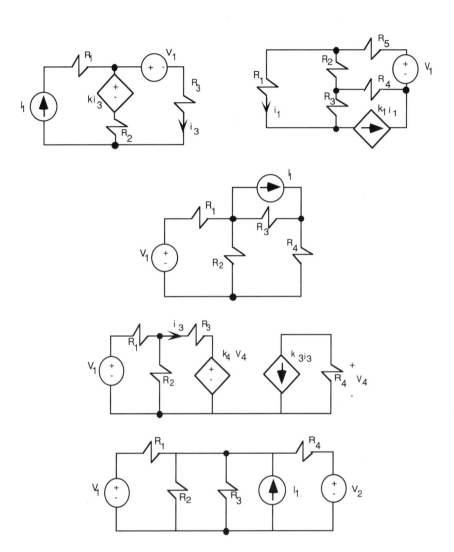

Figure 6.12. Circuits for problems (cont).

7

ENERGY-STORAGE ELEMENTS

Sources *generate* energy and resistors *dissipate* energy as heat. None of the circuit elements we have discussed so far can *store* energy. Energy-storage elements are necessary in many practically useful circuits. To produce a spark in a spark plug, or to flash a flash-bulb, it is necessary to store electrical energy for sudden release. Two energy-storage elements are discussed in this chapter: a) the *capacitor*, and b) the *inductor*. In a capacitor, energy is stored in an electrical field, and in an inductor, this is done in a magnetic field. However, for circuit analysis it is sufficient to know the terminal voltage/current relationships of these devices. Detailed knowledge of field theory is not essential.

7.1 Capacitors

Capacitors in circuits are symbolized by the letter **C** and their graphic symbol is shown in Figure 7.1. In a capacitor, two electrical conductors are separated by a *dielectric* or insulator. The graphic symbol in Figure 7.1 shows this separation between the conductors and does not imply an open circuit. Electrical charge can not be transported through an insulator in the same manner as through a conductor. However, a time-varying voltage between the terminals of a capacitor causes a time-varying displacement of charge within the insulator. This *displacement current* is indistinguish-able at the terminals from conduction current.

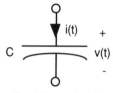

Figure 7.1. Graphic symbol for a capacitor.

The unit for measuring capacitance is called a *Farad* (F) and one Farad is a very high value for a capacitor. In common circuits, the value of a capacitor is often in the range of microfarads (uF) or picofarads (pF). The terminal voltage/current relationship of a capacitor, shown in Figure 7.1.(a), is

$$i = C \ dv/dt, \tag{7.1.1}$$

where i and v are functions of time t. Integrating both sides of (7.1.1) we obtain

$$v(t) = v(t_0) + [\int_{t_0}^{t} i(x)dx \ ] \ / \ C. \tag{7.1.2}$$

In (7.1.2) t_0 is any arbitrary starting time. By definition, an instantaneous change in $v(t)$ is the limit of δ going to zero $|v(t+\delta)-v(t)|$. From (7.1.2) we see that an instantaneous change in $v(t)$ is zero. Hence, the *terminal voltage of a capacitor can not change instantaneously.* This result is often used to find distributions of voltages and currents in circuits containing capacitors. Note that the terminal current $i(t)$ of a capacitor can change instantaneously.

We can arrive at the same result geometrically. From (7.1.2) we observe that the instantaneous change in the voltage $v(t)$ is proportional to the area under the curve of $i(t)$ vs. t. When the length of the interval δ along the time axis goes to zero, the corresponding area also goes to zero.

7.2 SPICE Simulation of Capacitors

Capacitors can be simulated as easily using SPICE as resistors. However, (7.1.1) and (7.1.2) indicate that the terminal voltages and currents should vary with time in order to obtain interesting results. Only time-varying voltages produce displacement currents. A constant terminal voltage produces zero terminal current. Figure 7.2 shows the SPICE simulation model of a simple capacitor circuit (see Figure 7.3) illustrating (7.1.1). The element lines and the control lines are discussed below.

```
capv.cir  circuit with a single capacitor
* input is a voltage source (see Figure 7.2)
* with piece-wise linear output

vs            1     0       cv1(0 0 2 3 3 2 4 2 5 4 6 0)

*element line for a capacitor
c1            1     2       4UF  ic+0V

* dead source for monitoring i1 in c1
vd            2     0       dc     0V

* transient analysis computes values of circuit
*   variables as functions of time
.tran 1 6   uic

.probe v(1,0) i(vd)
.options nopage
.end
```

Figure 7.2. SPICE simulation model of a simple capacitor circuit.

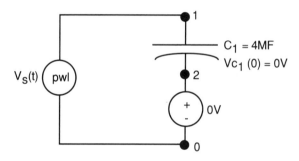

Figure 7.3. Simple capacitor circuit.

The input source to the circuit is a voltage source and its output is specified by a continuous *piece-wise linear* (**pwl**) function of time. Figure 7.4 shows the piece-wise linear function used in the circuit file in Figure 7.2. As the name indicates the function is linear over successive, nonoverlapping intervals of time and continuous at the boundary points. Each line segment is specified by its two end points each of which it shares with one another line segment on either side of it. The end points of each line are given as pairs of (T_k, V_k) where T_k is an instant of time and V_k is the corresponding voltage value. The voltage function shown in Figure 7.4 is given as **pwl**(0 0 2 3 3 2 4 2 5 4 6 0) in the circuit file.

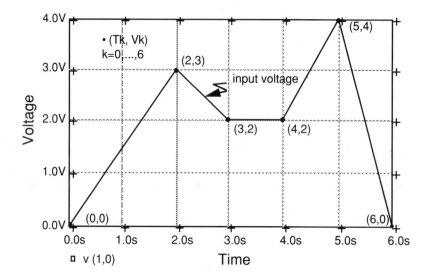

Figure 7.4. Piece-wise linear (**pwl**) function of time

The element line for the capacitor starts with its name **c1**. All capacitor names must start with the letter **c**. The value of the capacitor is 4 microfarad and it is entered on the element line as 4UF. The initial voltage on the capacitor is assumed to be zero. On the element line this fact is entered as **ic** = 0V.

The control line starts with the command **.tran**. This is an abbreviation for transient analysis. This type of analysis will be discussed in more detail in the next chapter. The output of such an analysis produces values of the circuit variables as functions of time. The most commonly used form of the transient analysis control line is **.tran tstep tstop**, which starts the transient analysis at time zero and ends at **tstop**. Note that **tstep** is not necessarily the time increment used in the simulation of transient behavior. The time increment is the smaller of **tstep** and **tstop**/50. **tstep** is the time increment used for plotting results of the analysis.The option **uic** indicates that the transient analysis should make use of the initial condition given in the element line.

The output of the simulation is the current through the capacitor and is shown in Figure 7.5. According to (7.1.1), the terminal current of a capacitor is proportional to the rate of change of the terminal voltage. Since the terminal voltage is **pwl** these rates are constants over the disjoint time intervals shown in Figure 7.4. The rates are 1.5, -1, 0, 2, and -4 volts/second respectively, and the corresponding current values according to (7.1.1) are 4 x (1.5, -1, 0, 2, -4) = (6, -4, 0, 8, -16) UA. These are the values for the current shown in Figure 7.5.

Figure 7.6 shows another SPICE simulation model of a circuit with two resistors, one capacitor, and a current source shown in Figure 7.7. The output of the current source goes through a very abrupt change of values between 4s and 4.01s. Otherwise it is **pwl**. The initial voltage on the capacitor is **ic** = -7V. According to (7.1.2) the voltage is proportional to t^2 since the current is **pwl**. The terminal current and the terminal voltage of the capacitor are shown in Figure 7.8 (a) and (b). The reader should verify that the terminal voltage/current relation satisfies (7.1.2). Note that the terminal voltage starts at the given initial value of -7V.

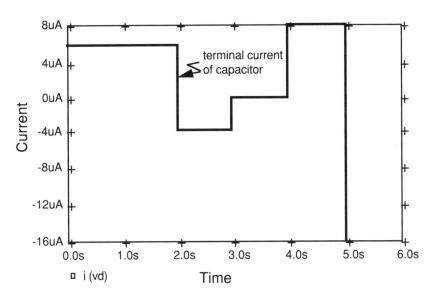

Figure 7.5. Current through capacitor.

```
cap.cir  circuit with a single capacitor
* input is a current source (consult Figure 7.4)

is1    0 1 pwl(0 0 2 20 3 0 4 0 4.01 -30 4.5 -30 4.51 0 6
              0)
r1 1   0 1MEG
c1 1   2 2F ic=-7V
r2 2   3 1
vd 3   0 dc          0V
.tran 1 6 uic
.probe v(1,2) i(vd)

.options nopage
.end
```

Figure 7.6. SPICE simulation model.

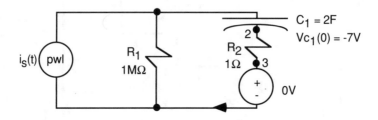

Figure 7.7. Circuit with two resistors, one capacitor, and a current source.

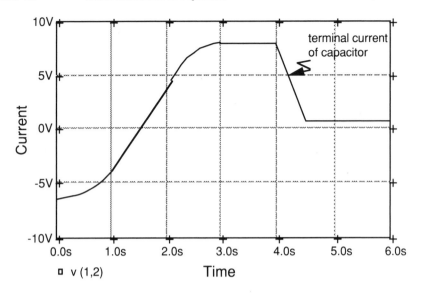

Figure 7.8 (a). Terminal voltage of capacitor.

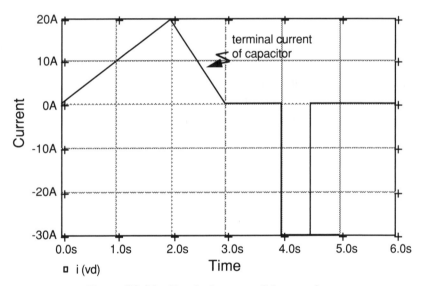

Figure 7.8 (b). Terminal current of the capacitor.

7.3 Inductors

Inductors in circuits are symbolized by the letter **L** and their graphic symbol is shown in Figure 7.9. In an inductor, an electrical conductor is wrapped around a magnetic material. The graphic symbol in Figure 7.9 shows such a spiralling conductor often called a coil. An electrical current in an inductor generates a magnetic field and produces a terminal voltage by electro-magnetic induction.

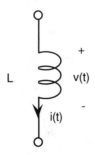

Figure 7.9. Graphic symbol for an inductor.

The unit for measuring inductance is called a *Henry* (H) and one Henry is a high value for an inductor except in electro-magnetic systems. In common circuits, the value of an inductor is often in the range of millihenrys (MH). Inductors are not directly used in most modern electronic circuits because of their bulk. Their effect is simulated by other means. The terminal voltage/current relationship of an inductor, shown in Figure 7.9, is

$$\mathbf{v} = \mathbf{L}\ di/dt, \qquad\qquad (7.3.1)$$

where **i** and **v** are functions of time t. Integrating both sides of (7.3.1) we obtain

$$\mathbf{i}(t) = \mathbf{i}(t_0) + [\int_{t_0}^{t} v(x)dx]\ /\ \mathbf{L} \qquad\qquad (7.3.2)$$

In (7.3.2) t_0 is any arbitrary starting time. By definition, an instantaneous change in $i(t)$ is the limit of δ going to zero $|i(t+\delta) - i(t)|$. From (7.3.2) we see that an instantaneous change in $i(t)$ is zero. Hence, *the terminal current of an inductor can not change instantaneously.* This result is often used to find distributions of voltages and currents in circuits containing inductors. Note that the terminal voltage $v(t)$ of an inductor can change instantaneously.

We can arrive at the same result geometrically. From (7.3.2) we observe that the instantaneous change in the current $i(t)$ is proportional to the area under the curve of $v(t)$ vs. t. When the length of the interval δ along the time axis goes to zero, the corresponding area also goes to zero.

7.4 SPICE Simulation of Inductors

Inductors can be simulated as easily using SPICE as resistors and capacitors. However, (7.3.1) and (7.3.2) indicate that the terminal voltages and currents should vary with time in order to obtain interesting results. Constant terminal currents produce zero terminal voltages. Figure 7.10 shows the SPICE simulation of a simple inductor (see Figure 7.11) illustrating (7.3.1). The element lines and the control lines are discussed below.

```
indv.cir  circuit with a single inductor
* input is a current source (consult Figure 7.11)

is   0    1     pwl (0 0 2 5 3 3 4 3 5 6 6 0)

* element line for inductor
l1   1    2     4MH ic = 0A

vd   2    0     dc    0V
.tran 1 6 uic
.probe v(1,2) i(vd)
.options nopage
.end
```

Figure 7.10. SPICE simulation model of a simple inductor.

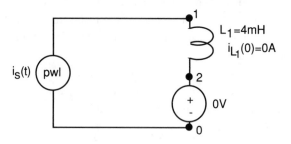

Figure 7.11. Circuit with simple inductor.

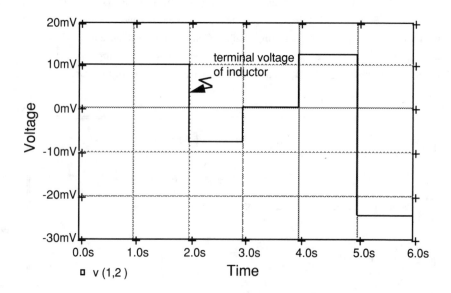

Figure 7.12 (a). Terminal voltage of inductor.

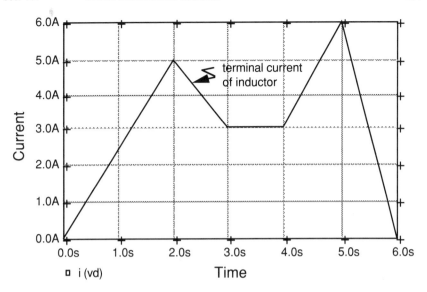

Figure 7.12 (b). Terminal current of inductor.

The output of the current source is taken to be a **pwl** function of time. The element line for the inductor starts with its name **l1**. All inductor names must start with the letter **l**. The value of the inductor is 4 millihenrys and it is entered on the element line as **4MH**. The initial current through the inductor is assumed to be zero. On the element line this fact is entered as **ic = 0A**. The control line **.tran** has already been discussed in Section 7.2.

Figure 7.13 shows another circuit containing an inductor simulated by SPICE. The output of the voltage source is a *pulse*. A pulse signal is described in SPICE as **pulse(V1 V2 TD TR TF PW PER)**. The parameters used in the description of a pulse are shown in Figure 7.14. The element line describes a voltage output **pulse(0 2 1 1 2 2 0)** which is shown in Figure 7.15. The corresponding current through the inductor is computed by (7.3.2) using the initial current value **ic = -1A** and is shown in Figure 7.16.

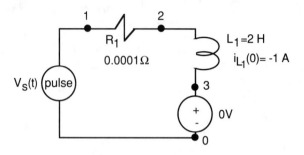

Figure 7.13. Circuit containing an inductor.

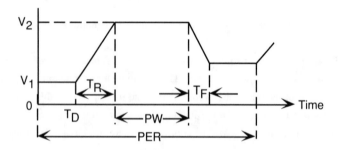

Figure 7.14. Pulse signal in SPICE.

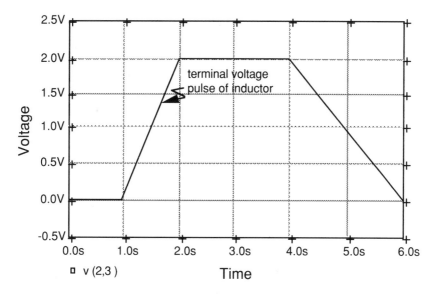

Figure 7.15. Voltage pulse input.

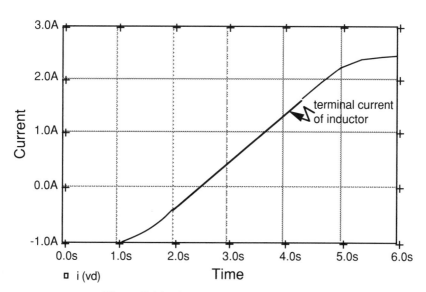

Figure 7.16. Current through the inductor.

The terminal current of the inductor starts at the initial value of -1A. As long as the voltage pulse is zero (from 0 to 1s), the current is zero. As the voltage increases linearly from 1s to 2s, the current increases as a function of t^2. When the voltage becomes constant (2V) over 2s to 4s, the current increases linearly over that period. The reader should be able to analyze the remaining portion of Figure 7.16 following (7.3.2).

7.5 Expressions for Energy

Since capacitors and inductors store energy we need expressions for energies stored in these elements. These expressions are derived from the basic relationships among energy $E(t)$, power $p(t)$, voltage $v(t)$, and current $i(t)$. For the capacitor we proceed as follows:

$$dE(t)/dt = p(t)$$
$$= v(t)\ i(t)$$
$$= C\ v(t)\ dv(t)/dt, \tag{7.5.1}$$

where the right hand side of (7.1.1) is substituted for $i(t)$. Integrating both sides of (7.5.1) and assuming that the reference for zero energy corresponds to zero voltage, we obtain the following expression of energy in a capacitor with terminal voltage $v(t)$:

$$E(t) = C\ v(t)^2\ /\ 2. \tag{7.5.2}$$

In a similar manner, using (7.3.1) we can show that the energy in an inductor with terminal current $i(t)$ is given by

$$E(t) = L\ i(t)^2\ /\ 2. \tag{7.5.3}$$

7.6 Response to Sinusoids

A sine or a cosine function of time is commonly called a *sinusoid*. Examples of a voltage and a current sinusoid are $v(t) = V \sin(\omega t)$ and $i(t) = I \cos(\omega t)$. The constants V and I are called the *amplitudes* of the corresponding sinusoids. The frequency constant ω

is called the *angular frequency* and is measured in radians per second. Two typical sinusoids are shown in Figure 7.17.

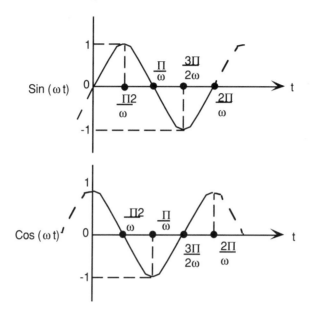

Figure 7.17. Two typical sinusoids.

Let the terminal voltage of a capacitor be given by $v(t) = \mathbf{V} \sin(\omega t)$. The terminal current $i(t)$ from (7.1.1) becomes $i(t) = \mathbf{V} \omega C \cos(\omega t) = \mathbf{V} \omega C \sin(\omega t + \pi / 2)$. The amplitude of the current sinusoid is $\mathbf{I} = \omega C \mathbf{V}$. Comparing $v(t) = \mathbf{V} \sin(\omega t)$ and $i(t) = \mathbf{I} \sin(\omega t + \pi / 2)$ we see that the current sinusoid has a 90 degrees phase lead over the voltage sinusoid. Compared to the voltage sinusoid, the current sinusoid leads in time by $\pi / 2\omega$ seconds. Figure 7.18 shows the terminal voltage/current relationship of a capacitor with sinusoidal input.

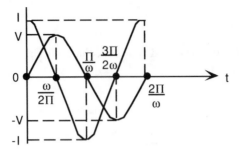

Figure 7.18. Terminal voltage/current relationship of capacitor with sinusoidal input.

The terminal voltage/current relationship of an inductor can be obtained in a similar manner using (7.3.1). Given $i(t) = I \sin(\omega t)$, we get $v(t) = I \omega L \sin(\omega t + \pi / 2)$. In case of an inductor, the terminal voltage sinusoid leads the current sinusoid by 90 degrees. In other words the terminal current sinusoid has a 90 degrees *phase lag* compared to the voltage sinusoid.

Figure 7.19 shows the SPICE simulation of a capacitor circuit with sinusoidal input (see Figure 7.20). A sinusoid in SPICE is described as **sin(V0 VA F TD 0)** where the significance of the parameters are shown in Figure 7.21. Note that the terminal current *leads* the terminal voltage by 1.5s. The time period of the sinusoid from the output of Figure 7.22 is **T** = 6s. By direct computation time lead = $\pi / 2\omega = \pi \, \mathbf{T} / 4 \, \pi$ = 1.5s.

```
capsin.cir   acapacitor circuit with
*  a sinusoidal voltage source (consult Figure 7.19)

vs    1     0      sin (1 1 0.167 0 0)
c1    1     2      2F
vd    2     0      dc    0v

.tran 1 6
.probe v(1,2) i(vd)
.options nopage
.end
```

Figure 7.19. SPICE simulation of a capacitor circuit with sinusoidal input.

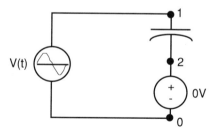

Figure 7.20. Capacitor circuit with sinusoidal input.

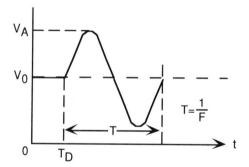

Figure 7.21. Significance of SPICE parameters used to describe sinusoid.

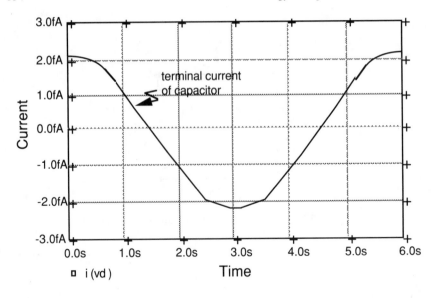

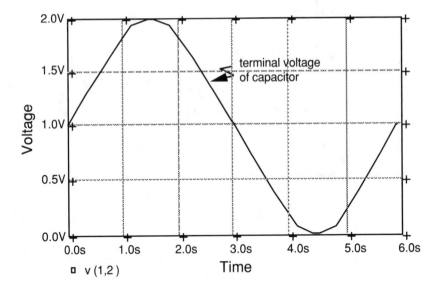

Figure 7.22. SPICE output showing the time period.

7.7 Response to Exponentials

An exponential function of time has the form exp(αt) where the exponent α is a real number. Figure 7.23 shows two possible types of exponential functions. The growth or decay in the value of an exponential is extremely rapid. The reciprocal of α is often called a *time-constant* of the exponential. A smaller time-constant means a faster growth or decay of the values of an exponential.

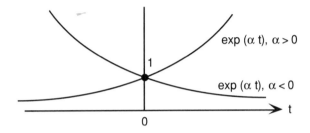

Figure 7.23. Two types of exponential functions.

Let the terminal voltage of a capacitor be $\mathbf{v}$(t) = **V** exp(αt). From (7.1.1), the terminal current becomes $\mathbf{i}$(t) = **V** α **C** exp(αt). Using (7.3.1) similar terminal voltage/current relationship for an inductor can be easily obtained.

Figure 7.24 shows the SPICE simulation of a capacitor with an exponential voltage source as input. The circuit is the same as shown in Figure 7.20 with the sinusoidal source replaced by an exponential source. An exponential source output is described in SPICE by **exp(V1 V2 TD1 T1 TD2 T2)** where the parameters are shown in Figure 7.25.

```
capexp.cir a capacitor circuit with
* an exponential voltage source

vs   1      0       exp (5 1 0 1.8 6 0)
c1   1      2       2F
vd   2      0       dc      0V

.tran 1 6
.probe v(1,2) i(vd)
.options nopage
.end
```

Figure 7.24. SPICE simulation of capacitor with exponential voltage source as input.

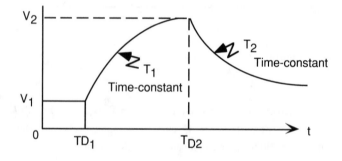

Figure 7.25. Significance of parameters used to describe exponential source output.

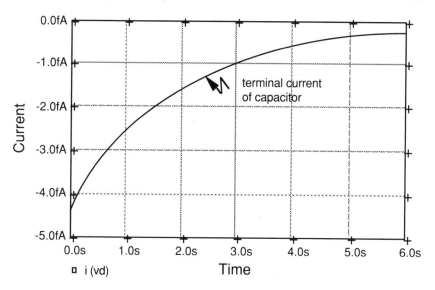

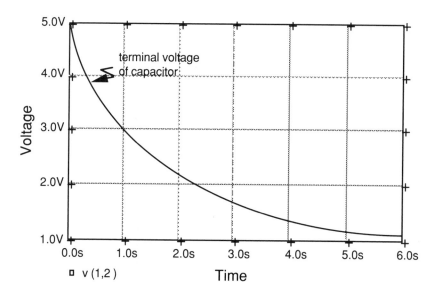

Figure 7.26.

7.8 Series and Parallel Combinations

Two capacitors in parallel are shown in Figure 7.27. By (7.1.1) $i_k(t) = C_k \, dv_k/dt = C_k \, dv/dt$, for $k = 1, 2$, and by Kirchhoff's current law $i(t) = i_1(t) + i_2(t)$. Combining these two equations we have, $i(t) = (C_1 + C_2) \, dv/dt$ and hence, the equivalent capacitance of two capacitors in parallel is given by the expression

$$C_{eq} = C_1 + C_2. \qquad (7.8.1)$$

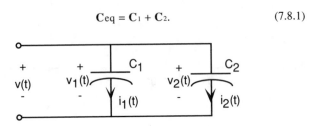

Figure 7.27. Capacitors in parallel.

Two capacitors in series are shown in Figure 7.28. Each capacitor C_k has an initial voltage $v_k(0)$ on it where we have assumed starting time $t_0 = 0$. By (7.1.2) $v_k(t) = v_k(0) + [\int_0^t i_k(x)dx] / C_k = v_k(0) + [\int_0^t i(x)dx] / C_k$ for $k = 1, 2$ and by Kirchhoff's voltage law $v(t) = v_1(t) + v_2(t)$. Combining these two equations we have, $v(t) = v_1(0) + v_2(0) + (1/C_1 + 1/C_2) [\int_0^t i_k(x)dx]$ and hence, the equivalent capacitance of two capacitors in series is given by the expression

$$C_{eq} = C_1 C_2 / (C_1 + C_2). \qquad (7.8.2)$$

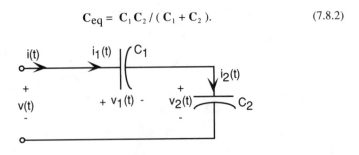

Figure 7.28. Capacitor in series.

The initial voltage $v_{eq}(0)$ on the equivalent capacitor is given by

$$v_{eq}(0) = v_1(0) + v_2(0). \qquad (7.8.3)$$

Expressions for the equivalent inductors of the series and the parallel combinations of inductors can be obtained in a similar way. These are (a) for the series combination

$$\mathbf{L_{eq} = L_1 + L_2}, \qquad (7.8.4)$$

and (b) for the parallel combination

$$\mathbf{L_{eq} = L_1 L_2 / (L_1 + L_2)}, \qquad (7.8.5)$$

and

$$i_{eq}(0) = i_1(0) + i_2(0). \qquad (7.8.6)$$

7.9 Examples

Example 7.9.1

(*Singular case*) Consider the circuit shown in Figure 7.29. The capacitor C_1 has an initial voltage of $v_1(0^-) = v_1(0) = 1V$. The initial energy stored in C_1 is $E_1(0^-) = E_1(0) = C_1 v_1(0)^2 / 2 = 1/2$ joule. In a similar manner, the initial energy stored in C_2 is $E_2(0^-) = E_2(0) = 0$ J. The combined energy in the system is $E(0) = E_1(0) + E_2(0) = 1/2$ J.

After the switch S is closed at t=0, the two capacitors are in parallel and they have a common terminal voltage of $v(t)$ (see Figure 7.30). The node equation becomes $C_1 \, dv/dt + C_2 \, dv/dt = 0$. Integrating both sides of this expression with respect to time over [0, 0^+] we get $C_1 [v(0^+) - v_1(0)] + C_2 [v(0^+) - v_2(0)] = C_1 [v(0^+) - 1] + C_2 [v(0^+) - 0] = 0$, or $v(0^+) = 1/2$. The combined energy in the system is $E(0^+) = (C_1 + C_2) v(0^+)^2 / 2 = 1/4$ J.

Note that the voltages across the capacitors are changing instantaneously by 1/2 V and 1/4 J of energy is missing although the amount of electrical charge in the system remains constant at 1 Q. Our model of a capacitor given by (7.1.1) or (7.1.2) is no longer valid here over 0^- to 0^+. When the switch is closed at t=0, a large current is produced as charge is transferred from C_1 to C_2. This rapid change in current over a short period of time generates an electromagnetic wave and radiates 1/4 J of energy. The terminal voltages change over a very short but nonzero interval of time.

Example 7.9.2

(Duality) Compare equations (7.1.1) and (7.3.1). If **v** and **i** are interchanged in (7.1.1) and **C** is replaced by **L**, then we obtain (7.3.1). A similar transformation takes us from (7.3.1) to (7.1.1). Because of this transformation, capacitors and inductors are called duals of each other. Valid results for one circuit can be transformed into valid dual results for a dual circuit. An example is given by the expressions of energy (7.5.2) and (7.5.3).

Example 7.9.3

Derive expressions for power and energy in a capacitor whose terminal voltage is a sinusoid $v(t) = V \sin(\omega t)$.

Given that $v(t) = V \sin(\omega t)$ is the terminal voltage of a capacitor **C**, from Section 7.6 we have $i(t) = wCV \cos(\omega t)$ and power $p(t) = v(t) i(t) = \omega CV^2 \sin(\omega t) \cos(\omega t) = \omega CV^2/2 \sin(2\omega t)$. Clearly, power $p(t)$ is also a sinusoid but at twice the frequency of the terminal voltage. Figures 7.31 and 7.32 show the terminal voltage, the terminal current and the power as functions of time.

Since power is given by a sinusoid, average power over a period of π / ω is zero and the same is true for the energy. Over half a period $(\pi / 2\omega)$ average energy (by integration of power) equals CV^2 / π.

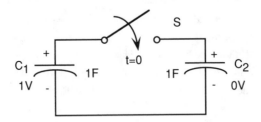

Figure 7.29.

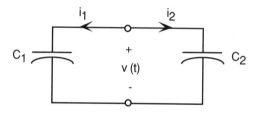

Figure 7.30.

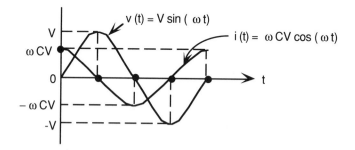

Figure 7.31.

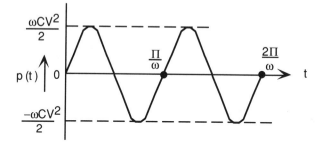

Figure 7.32.

7.10. Summary

- Capacitors store electrical energy in electrical fields. Inductors store electrical energy in magnetic fields.

- Terminal voltage/current relations for a capacitor and an inductor are $i(t) = C\, dv/dt$ and $v(t) = L\, di/dt$ respectively.

- The terminal voltage of a capacitor and the terminal current of an inductor can not change instantaneously.

- The energy at time t in a capacitor (inductor) is given by $C\, v(t)^2/2$ ($L\, i(t)^2/2$).

- With sinusoidal input, the terminal current of a capacitor leads its terminal voltage by $\pi / 2$.

- With sinusoidal input, the terminal voltage of an inductor leads its terminal current by $\pi / 2$.

- The formulas for the series and the parallel combinations of two capacitors are $C_{eq} = C_1 C_2 / (C_1 + C_2)$ and $C_{eq} = C_1 + C_2$ respectively.

- The formulas for the series and the parallel combinations of two inductors are $L_{eq} = L_1 + L_2$ and $L_{eq} = L_1 L_2 / (L_1 + L_2)$ respectively.

- In SPICE, transient analysis can be done by the **.tran** control line. Possible time-varying input functions are **pwl**, **pulse**, **sin**, and **exp**. Consult appropriate sections of this chapter for details.

7.11 Problems

7.1.   Derive (7.5.3) for an inductor.

7.2.   Given $i(t) = I\, \sin(\omega t)$, using (7.1.2) derive an expression for $v(t)$. Check the lead/lag condition between $i(t)$ and $v(t)$.

7.3.   Given $v(t) = V \sin(wt)$, using (7.3.2) derive an expression for $i(t)$. Check the lead/lag condition between $v(t)$ and $i(t)$.

7.4.   Using SPICE simulate a circuit with an inductor and a sinusoidal source. Plot power in the inductor as a function of time. Derive an expression for the power in the inductor and compare with the simulated result.

7.5.   Using SPICE simulate a circuit with an inductor and an exponential source. Derive expressions for the power and the energy in the inductor.

7.6.   Two capacitors C_1 and C_2 with initial voltages v_1 and v_2 and energies E_1 and E_2 are connected in series to form an equivalent capacitor C_{eq} with initial energy E_{eq}. Show that if $C_1 v_1 = C_2 v_2$, then $E_{eq} = E_1 + E_2$.

7.7.   Derive (7.8.4), (7.8.5), and (7.8.6).

7.8.   Derive the current and the voltage division formulas for capacitors.

7.9.   Derive the current and the voltage division formulas for inductors.

7.10.   Perform a SPICE simulation of a circuit where a 2F capacitor with **ic** = 4V is connected across a 2H inductor with **ic** = 0A. Use **.tran 0.1Us 0.6Us   uic**. Plot the voltages and currents in the circuit. From physical considerations explain the time-varying nature of these circuit variables.

8

NATURAL RESPONSES OF RC AND RL CIRCUITS

Capacitors and inductors are energy-storage elements. When the stored energy is released, the values of the circuit variables change with time. In this chapter we shall analyze these changes for two special types of circuits, called the RC and the RL respectively. A RC circuit is made up only of resistors and capacitors. Furthermore, in a RC circuit, all the resistors and all the capacitors can be separately combined into a single equivalent resistor and a single equivalent capacitor respectively. Actually, because of the Thevenin and Norton theorem, it is enough to require that all capacitors be combined into a single equivalent capacitor. The resistive part of the circuit can be replaced by its Thevenin or Norton equivalent circuit. RL circuits have similar properties except that inductors are used instead of capacitors.

We also assume that when the stored energy is released from a capacitor or an inductor, there are no other sources present in the circuit. The behaviors of the circuit variables are controlled by the structure of the circuit alone, and are not influenced by external sources. In this sense, the response of the circuit to the energy released is a *natural response*.

8.1 Analysis from Physical Considerations

Consider a RC circuit such as the one shown in Figure 8.1. S represents a switch which is closed at $t = 0$. Before the closing of the switch the circuit is open and hence, the current in the circuit is zero. The voltage across the capacitor is V_0 before the switch is closed and therefore the initial energy stored in the capacitor is $C V_0^2 / 2$. In this section we find the two limiting values of all the circuit variables from physical considerations.

For each circuit variable we are interested in two limiting values. One is at $t = 0^+$ where 0^+ is positive and infinitely close to 0. The other limiting value is attained as t approaches infinity. Consult Figure 8.2 for the symbols used in the following discussion.

Since the voltage across a capacitor can not change instantaneously, $v_C(0^+) = v_C(0) = V_0$. By Kirchhoff's voltage law $v_R(0^+) = v_C(0^+) = V_0$, and by Ohm's law $i_R(0^+) = V_0 / R$. These are the initial values of the circuit variables at $t = 0$.

As current flows in the circuit, the resistor dissipates energy at the rate given by $R i(t)^2$. Since the initial energy stored in the capacitor is finite, eventually all of it is dissipated. Hence, as time approaches infinity all the values of the circuit variables approach zeros. Now that we know the two limiting values of every circuit variable entirely from physical considerations, we proceed to find out how they change from one to the other.

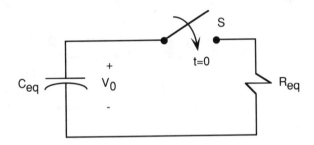

Figure 8.1. RC circuit.

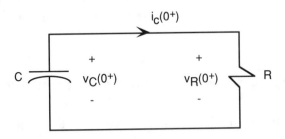

Figure 8.2.

8.2 Analysis of a RC Circuit

In any circuit, Kirchhoff's voltage and current laws are satisfied at all times. Applying Kirchhoff's voltage law to the circuit in Figure 8.2 we get

$$-v_C(t) + v_R(t) = 0. \qquad (8.2.1)$$

From (7.1.1) we know

$$i_C(t) = -C\, dv_C(t)/dt. \qquad (8.2.2)$$

The minus sign is due to the fact that the direction of $i_C(t)$ in Figure 7.1 is opposite from that shown in Figure 8.2. Since by Ohm's law

$$v_R(t) = R\, i_C(t), \qquad (8.2.3)$$

combining (8.2.1), (8.2.2), and (8.2.3) we get

$$dv_C(t)/dt + v_C(t)\,/\,RC = 0 \qquad (8.2.4)$$

The expression in (8.2.4) is an equation in the unknown function $v_C(t)$. Since (8.2.4) involves the first derivative of $v_C(t)$ and differentiation is a linear operation, (8.2.4) is called a first-order linear differential equation. Once (8.2.4) is solved for the unknown $v_C(t)$, the other circuit variable $i_C(t)$ can be obtained from (8.2.2).

An equation for the unknown $i_C(t)$ can be obtained independent of $v_C(t)$. From (7.1.2) we know

$$v_C(t) = -V_0 - [\int_0^t i_C(x)dx]\,/C \qquad (8.2.5)$$

Substituting (8.2.5) in (8.2.1) and using Ohm's law we get

$$V_0 + [\int_0^t i_C((x)dx] / C + R\ i_C(t) = 0. \tag{8.2.6}$$

Now differentiating both sides of (8.2.6) with respect to time we have

$$di_C(t)/dt + i_C(t) / RC = 0 \tag{8.2.7}$$

Again we obtain a first-order linear differential equation for the unknown $i_C(t)$. The solution of a first-order linear differential equation is derived in the next section.

8.3   First-Order Linear Differential Equation

The general form of (8.2.4) or (8.2.7) is

$$dx(t)/dt + x(t) / T = 0, \tag{8.3.1}$$

where $x(t)$ is an unknown function of time. Solving (8.3.1) is extremely simple and requires knowledge only of differential calculus. We can rewrite (8.3.1) as

$$dx(t)/dt = -x(t) / T . \tag{8.3.2}$$

From (8.3.2) it is clear that $x(t)$ is a function such that its first derivative has the same form as $x(t)$ itself. From differential calculus we know that the exponential function $A \exp(\alpha t)$ is such a function. Setting $\alpha = -1/T$, we take $x(t) = A \exp(-t/T)$ which obviously satisfies (8.3.2) and also (8.3.1).

The solution of (8.3.1) is $x(t) = A \exp(-t/T)$ where A is an unknown constant. Thus we have a family of solutions to (8.3.1). A specific value of A picks out a particular member of this family. For example if $x(0) = X_0$ is given, then $x(t) = X_0 \exp(-t/T)$ becomes a particular solution satisfying the given initial condition $x(0) = X_0$.

Figure 8.3 shows the general shape of $x(t) = A \exp(-t/T)$ for positive values of T. This curve of $x(t)$ is known in general as a decaying exponential. From (8.3.2) we know that the rate of change of x with time is controlled by the value of T. The smaller this value of T, the faster is the change in x. T is called a *time-constant* and the response $x(t)$

becomes close to zero within a few multiples of the time-constant. After one time-constant $x(t) = 0.368$ A approximately.

According to (8.3.2), a tangent drawn to $x(t)$ at $t = 0$ has the slope of $-A/T$. The tangent line obviously passes through the point $(0,A)$. With a slope of $-A/T$, the tangent line must intersect the time axes at $t = T$.

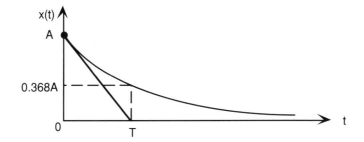

Figure 8.3. Decaying exponential function.

8.4 Circuit Variables in a RC Circuit

In Section 8.2 we found that the circuit variables in a RC circuit satisfy first-order linear differential equations (8.2.4) and (8.2.7). The general solution of such an equation is derived in Section 8.3 as A exp($-t/T$). The initial values of the circuit variables are derived in Section 8.1. Putting all these together with $T = RC$ we have

$$v_C(t) = V_0 \exp(-t/RC), \tag{8.4.1}$$

and

$$i_C(t) = (V_0/R) \exp(-t/RC). \tag{8.4.2}$$

The time-constant of a **RC** circuit is given by the product **RC** and it controls the rate of decay of the natural response.

8.5 SPICE Simulation of a RC Circuit

SPICE can be used to simulate the natural responses of a whole family of RC circuits. The effect of the time-constant on the rate of decay of the natural response becomes clear at a glance from the graphical output of such a simulation. Figure 8.4 shows the circuit file of a family of RC circuits simulated simultaneously. The natural responses are plotted on the same pair of axes in Figure 8.5. The time-constants are 2, 4, 8, and 16 milliseconds respectively. Clearly the result of the simulation shows that, higher time-constant values lead to slower rate of decay.

```
rcnat.cir natural response of rc circuits
*four RC circuits are run in parallel to
*observe effects of variations of time-constant
*value from 2MS to 16 MS

r1   0    1      2K
c1   1    0      0.001M        ic = 4V

r2   0    2      4K
c2   2    0      0.001M        ic = 4V

r3   0    3      8K
c3   3    0      0.001M        ic = 4V

.tran   0.002   0.06   uic
.probe   v(1)   v(2)   v(3)   v(4)
.options   nopage
.end
```

Figure 8.4. Circuit file of a family of RC circuits simulated simultaneously.

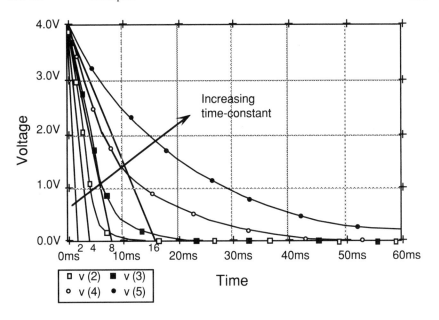

Figure 8.5. Natural response of a RC circuit.

8.6 Examples

Example 8.6.1

Find the natural response of the RC circuit shown in Figure 8.6 (a). The circuit shown has only one capacitor but four resistors. However, by simple series and parallel combination, the four resistors can be replaced by an equivalent resistor $R_{eq} = (R_1 + R_2)$ $(R_3 + R_4) / (R_1 + R_2 + R_3 + R_4)$ as shown in Figure 8.6 (b). This equivalent circuit is similar to the circuit shown in Figure 8.1 and can be analyzed in a similar manner.

The current in the branch containing R_1 and R_2 will in general be different from that in the other branch containing R_3 and R_4. Once the terminal voltage $v_C(t)$ of the capacitor is known, these currents can be obtained by Ohm's law. Alternately, once the terminal current $i_C(t)$ is known, the branch currents can be obtained by current division.

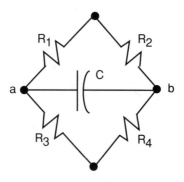

Figure 8.6 (a).

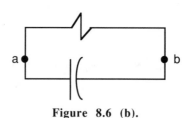

Figure 8.6 (b).

Example 8.6.2

Analyze the circuit shown in Figure 8.7. This circuit may look similar to the one shown in Figure 7.29, however, note the presence of the resistor. Because of the resistor, the current in the circuit can not reach infinity and no singular behavior is exhibited here.

This again is a simple RC circuit where the two capacitors in series can be replaced by an equivalent capacitor with an equivalent initial voltage. Once that is done, expressions derived earlier can be applied directly. The terminal voltages of the individual capacitors can then be obtained by integrating the expression of the common terminal current.

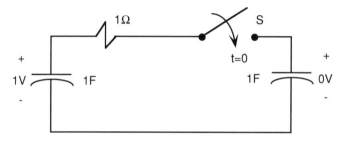

Figure 8.7.

8.7 Summary

- A RC circuit is made up only of resistors and capacitors. Furthermore, in a RC circuit, all the resistors and all the capacitors can be separately combined into a single equivalent resistor and a single equivalent capacitor respectively. RL circuits have similar properties except that inductors are used instead of capacitors.

- The natural response of a RC or RL circuit is the response of the circuit to the energy released by the capacitor or the inductor in the absence of other sources.

- The natural response of a RC or RL circuit satisfies a first-order linear differential equation.

- The natural response of a RC (RL) circuit is a decaying exponential with a time-constant given by **RC (L/R)**.

- The time-constant of a decaying exponential controls the rate at which the exponential decays. The smaller the time-constant, the faster is the rate of decay.

8.8 Problems

8.1.   Consider the RL circuit shown in Figure 8.8 where the switch S is moved from position a to position b at t = 0. What are the limiting values of the circuit variables?

8.2. Derive the first-order linear differential equation satisfied by $v_L(t)$ in the circuit shown in Figure 8.8.

8.3. Derive the first-order linear differential equation satisfied by $i_L(t)$ in the circuit shown in Figure 8.8.

8.4. What is the time-constant of a RL circuit? Sketch its natural response as a function of time.

8.5. Simulate a family of RL circuits using SPICE, and plot their natural responses on the same pair of axes. From the plot, estimate the time-constant of each circuit simulated and compare against the corresponding values obtained from the circuit file.

8.6. The voltages across the parallel combination of capacitors C_1 and C_2 at t=0 is V_0 in the circuit shown in Figure 8.9. The switch S is closed at t=0. Find the expressions for the currents in the capacitors after the switch is closed. Test the correctness of the derived expressions via. SPICE simulation.

8.7. For the circuit shown in Figure 8.10, the switch S is moved from a to b at t=0. Find expressions for the voltages across the two inductors after t=0. Test the correctness of the derived expressions via. SPICE simulation.

8.8. Find an expression for the time-constant of the RC circuit shown in Figure 8.11.

8.9. Find an expression for the time-constant of the RL circuit shown in Figure 8.12.

8.10. Find expressions for power and energy in a RC circuit.

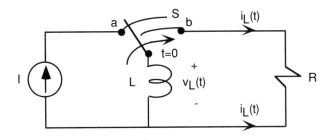

Figure 8.8.

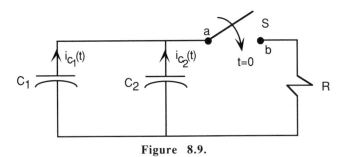

Figure 8.9.

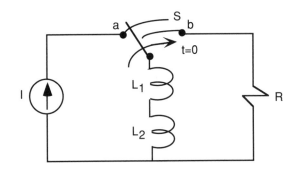

Figure 8.10.

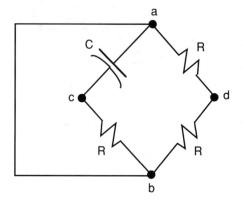

Figure 8.11.

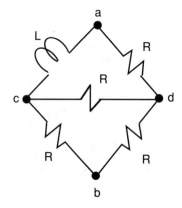

Figure 8.12.

9

STEP-RESPONSES OF RC AND RL CIRCUITS

Capacitors and inductors are energy-storage elements. Energy-storage elements are necessary in many practically useful circuits. To produce a spark in a spark plug, or to flash a flash-bulb, it is necessary to store electrical energy for sudden release. In Chapter 8 we discussed the behavior of RC and RL circuits when electrical energy is suddenly released. Now we consider the process of storing energy in capacitors and inductors in RC and RL circuits.

Energy is transferred into capacitors and inductors by connecting them across constant sources. If this connection is made at time t=0, the input to the circuit goes through a step change at t=0. The response of a RC or a RL circuit to such a step change of input is called its *step-response*.

9.1 Step-Response of a RC Circuit

9.1.1 Input Voltage Source

Consider the RC circuit with a constant voltage source shown in Figure 9.1. The switch S is closed at t=0 when the initial voltage across the capacitor is $v_C(0) = V_0$. Since the terminal voltage of a capacitor can not change instantaneously, $v_C(0^+) = V_0$. Hence, by Kirchhoff's voltage law $v_R(0^+) = V_S - v_C(0^+) = V_S - V_0$, and by Ohm's law $i(0^+) = (V_S - V_0) / R$.

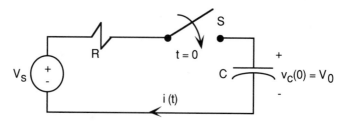

Figure 9.1. RC circuit with a constant voltage source.

The current in the RC circuit causes electrical charge to be transferred from the source to the capacitor. As the capacitor charges up its terminal voltage increases and eventually equals that of the source. By Kirchhoff's voltage law the terminal voltage of the resistor goes to zero. Hence, by Ohm's law the current in the circuit also goes to zero.

For time t greater than zero, the current in the circuit $i(t) = C \, dv_C/dt$. Using Kirchhoff's voltage law we can write

$$RC \, dv_C/dt + v_C(t) = V_S,$$

or

$$dv_C/dt + v_C(t)/RC = V_S/RC. \tag{9.1.1}$$

This is a first-order linear differential equation in the unknown function $v_C(t)$. However, the right hand side (RHS) of (9.1.1) is nonzero. The method of solution of this type of equation will be discussed in the next section.

Using (7.1.2) and Kirchhoff's voltage law we can also write

$$R \, i(t) + V_0 + [\int_0^t i(x)dx \,] / C = V_S. \tag{9.1.2}$$

Differentiating both sides of (9.1.2) with respect to time we have

$$di(t)/dt + i(t)/RC = 0. \tag{9.1.3}$$

We have solved equations like (9.1.3) in Chapter 8. The solution is $i(t) = A \exp(-t/RC)$, where $A = i(0)$. We have already derived an expression for $i(0)$. Hence, the complete solution is $i(t) = (V_S - V_0)/R \exp(-t/RC)$.

9.1.2. Input Current Source

The RC circuit shown in Figure 9.2 uses a current source as input. Before the switch S is closed at t=0, the current I_S from the source flows through the resistor **R**. By Ohm's law, the terminal voltage of the resistor becomes $v_R(0^-) = R\,I_S$. Since the terminal voltage of a capacitor can not change instantaneously, $v_C(0^+) = V_0 = v_R(0^+)$. By Ohm's law, $i_R(0^+) = V_0/R$. By Kirchhoff's current law, $i_C(0^+) = I_S - i_R(0^+) = I_S - V_0/R$.

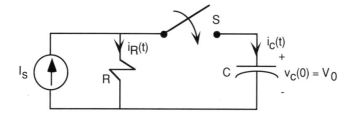

Figure 9.2. RC circuit with current source as input.

As the capacitor charges up its terminal voltage increases. Since it is in parallel with the resistor, the terminal voltage of the resistor also increases. By Ohm's law the current through the resistor increases. By Kirchhoff's current law, less and less current flows through the capacitor. Eventually the current in the capacitor goes to zero and its terminal voltage becomes $R\,I_S$.

For time t greater than zero, using Kirchhoff's current law we can write

$$C\,dv_C/dt + v_C(t)/R = I_S. \qquad (9.1.4)$$

The solution of this equation is discussed in the next section. Using (7.1.2) and Kirchhoff's current law we can also write

$$i_C(t) + [\, V_0\, (\int_0^t i_C(x)dx\,)/C\,]/R = I_S. \qquad (9.1.5)$$

Differentiating both sides of (9.1.5) with respect to time we have

$$di_C/dt + i_C(t)/RC = 0. \qquad (9.1.6)$$

We have solved equations like (9.1.6) in Chapter 8. The solution is $i_C(t) = A \exp(-t/RC)$, where $A = i_C(0)$. We have already derived an expression for $i_C(0)$. Hence, the complete solution is $i_C(t) = (I_S - V_0/R) \exp(-t/RC)$.

9.2   First-Order Linear Differential Equation

The general form of the first-order linear differential equation in (9.1.1) and (9.1.4) is

$$dx/dt + x(t)/T = K, \qquad (9.2.1)$$

where K is a constant. To solve (9.2.1) we assume that the solution is made up of two parts: a) the *transient solution*, and b) the *steady-state solution*. The total solution is the sum of these two parts.

Transient solution: We assume that the transient solution is the result of nonzero initial value of $x(t)$ and any discontinuity in the input. It is called a transient solution because it decays with time like the natural response. To find the transient solution we solve

$$dx/dt + x(t)/T = 0. \qquad (9.2.2)$$

We have already solved this equation in Chapter 8 and the solution is $x(t) = A \exp(-t/T)$. As before T is the time-constant of the circuit.

Steady-state solution: We assume that in the steady-state the function $x(t)$ has the same form as the input. In (9.2.1) the input is a constant K. Hence, in steady-state, the function $x(t)$ becomes some unknown constant B. Since the steady-state solution must also satisfy (9.2.1), we substitute B for $x(t)$ in (9.2.1) and obtain $B = K T$. The total solution becomes

$$x(t) = A \exp(-t/T) + K T. \qquad (9.2.3)$$

The unknown constant A is obtained by solving (9.2.3) for some known value of $x(t)$. For example if $x(0) = X_0$, then

$$x(t) = (X_0 - K\,T)\ \exp(\,-t/T\,) + K\,T.$$
$$= X_0 \exp(\,-t/T\,) + K\,T\,(\,1 - \exp(\,-t/T\,)) \qquad (9.2.4)$$

Figures 9.3 (a) and 9.3 (b) show possible shapes of $x(t)$ as a function of time. In Figure 9.2.(a), X_0 and K are assumed to be positive. In Figure 9.2.(b), X_0 is assumed to be zero, and K is still positive.

From the Figures 9.3 (a) and 9.3 (b) we see that the response $x(t)$ to a constant input K reaches a steady-state value of $K\,T$ and the speed of response is controlled by the time-constant T. The smaller the value of T, the quicker $x(t)$ approaches its steady-state value. After one time constant $x(t) = 0.632\ K\,T$ approximately.

From (9.2.1) we see that when $X_0 = 0$ the tangent to the $x(t)$ curve at t=0 has the slope of K. Hence, this tangent line intersects the constant KT line at $t = T$.

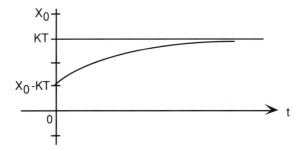

Figure 9.3 (a).

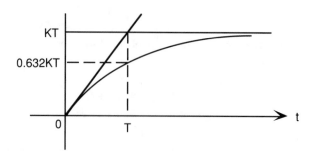

Figure 9.3 (b).

9.3   Circuit Variables in a RC Circuit

In Section 9.1 we found that the circuit variables in a RC circuit satisfy first-order linear differential equations (9.1.1) and (9.1.4). The general solution of such an equation is derived in Section 9.2 as A exp(-t/$\mathbf{T}$) + $\mathbf{K}$ $\mathbf{T}$. The initial values of the circuit variables are derived in Section 9.1. Putting all these together with $\mathbf{T} = \mathbf{RC}$ we have from (9.1.1)

$$v_C(t) = (\mathbf{V}_0 - \mathbf{V}_S) \exp(-t/\mathbf{RC}) + \mathbf{V}_S,$$

or

$$= \mathbf{V}_0 \exp(-t/\mathbf{RC}) + \mathbf{V}_S (1 - \exp(-t/\mathbf{RC})), \qquad (9.3.1)$$

and from (9.1.4)

$$v_C(t) = (\mathbf{V}_0 - \mathbf{R} \mathbf{I}_S) \exp(-t/\mathbf{RC}) + \mathbf{R} \mathbf{I}_S,$$

or

$$= \mathbf{V}_0 \exp(-t/\mathbf{RC}) + \mathbf{R} \mathbf{I}_S (1 - \exp(-t/\mathbf{RC})) \qquad (9.3.2)$$

The time-constant of a RC circuit is given by the product $\mathbf{RC}$ and it controls the rate of increase of the step-response.

9.4 SPICE Simulation of Step-Response

SPICE can be used to simulate the step-responses of a whole family of RC circuits. The effect of the time-constant on the rate of increase of the step-response becomes clear at a glance from the graphical output of such a simulation. Figure 9.4 shows the circuit file of a family of RC circuits simulated simultaneously. The natural responses are plotted on the same pair of axes in Figure 9.5. The time-constants are 2, 4, 8, and 16 milliseconds respectively. Clearly the result of the simulation shows that, higher time-constants lead to slower rates of increase.

```
steprc.cir step response of rc circuits
*time-constraints are 2, 4, 8, and 16 MS
*step height is 4V

r1   1    2     2K
c1   2    0     0.001M        ic = 0V

r2   1    3     4K
c2   3    0     0.001M        ic = 0V

r3   1    4     8K
c3   4    0     0.001M        ic = 0V

r4   1    5     16K
c4   5    0     0.001M        ic = 0V

vin  1    0        pulse (0 4 0 0 0 100M 100M)

.tran  0.002  0.0g  uic
.probe   v(2) v(3) v(4) v(5)
.options nopage
.end
```

Figure 9.4. Circuit file of a family of RC circuits simulated simultaneously.

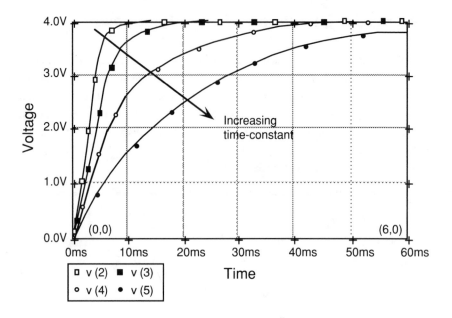

Figure 9.5. Natural response of a RC circuit.

9.5 Examples

Example 9.5.1

Analyze the RC circuit shown in Figure 9.6. An obvious approach to the analysis of this and other similar circuits is to replace the resistive part of the circuit including the source by its Thevenin or Norton equivalent circuit. The transformed circuits are shown in Figures 9.7 and 9.8. The expressions for V_{oc} and R_{eq} have already been derived in (5.7.4) and (5.7.5). The material of Section 9.1 can now be applied directly to the transformed circuits.

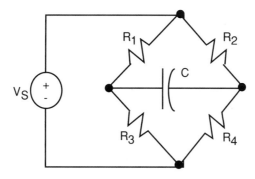

Figure 9.6.

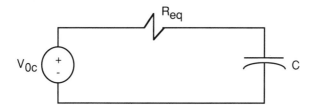

Figure 9.7.

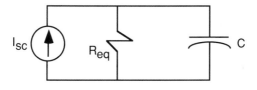

Figure 9.8.

Example 9.5.2

Analyze the circuit shown in Figure 9.9. Again by using Norton equivalent circuit in this case, the circuit is transformed into one shown in Figure 9.10. In this case $\mathbf{R}_{eq} =$

$R_1 + R_2$ and $I_{sc} = I_S (R_1 / (R_1 + R_2))$. The material from Section 9.1.2 can now be applied directly.

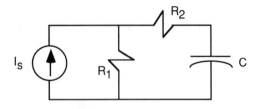

Figure 9.9.

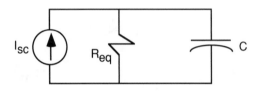

Figure 9.10.

9.6 Summary

- The step-response of a RC or a RL circuit is the response to a instantaneous change in the output of a source.

- The step-response satisfies a first-order linear differential equation with a constant nonzero RHS.

- The time-constant of the step-response of a RC (RL) circuit is given by **RC (L/R)**.

9.7 Problems

9.1.   Consider the RL circuit shown in Figure 9.11 where the switch S is moved from position a to position b at t = 0. What are the limiting values of the circuit variables?

9.2.   Derive the first-order linear differential equation satisfied by $v_L(t)$ in the circuit shown in Figure 9.11.

9.3.   Derive the first-order linear differential equation satisfied by $i_L(t)$ in the circuit shown in Figure 9.11.

9.4.   What is the time-constant of a RL circuit? Sketch its step-response as a function of time.

9.5.   Simulate a family of RL circuits using SPICE, and plot their step-responses on the same pair of axes. From the plot, estimate the time-constant of each circuit simulated and compare against the corresponding values obtained from the circuit file.

9.6.   The initial energy in the capacitor C in the circuit shown in Figure 9.12 is zero. The switch S is closed at t=0. Find the expressions for the terminal current and the terminal voltage of the capacitor after the switch is closed. What are the time-constant and the steady-state value of the terminal voltage of the capacitor? Test the correctness of the derived expressions via. SPICE simulation. (Hint: Use Thevenin's or Norton's equivalent circuit.)

9.7.   For the circuit shown in Figure 9.13, the initial energy in the inductor L is zero and the switch S is closed t=0. Find expressions for the terminal voltage and the terminal current of the inductor after t=0. Test the correctness of the derived expressions via. SPICE simulation. (Hint: Use Thevenin's or Norton's equivalent circuit.)

9.8.   Find an expression for the time-constant of the RC circuit shown in Figure 9.14. (Hint: Use Thevenin's or Norton's equivalent circuit.)

9.9.   Find an expression for the time-constant of the RL circuit shown in Figure 9.15. (Hint: Use Thevenin's or Norton's equivalent circuit.)

9.10.   Find expressions for power and energy in a RC circuit while the capacitor is being charged. (Hint: Use Thevenin's or Norton's equivalent circuit.)

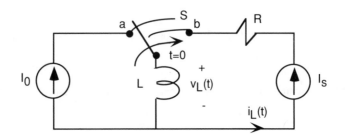

Figure 9.11.

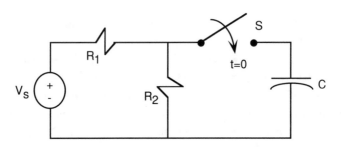

Figure 9.12.

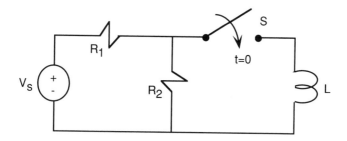

Figure   9.13.

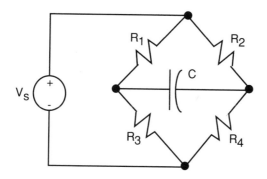

Figure   9.14.

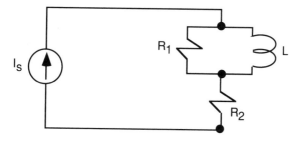

Figure   9.15.

10

RESPONSES OF RLC CIRCUITS

A RLC circuit contains capacitors as well as inductors. Since we have studied RC and RL circuits separately it may appear that a RLC circuit can be analyzed by borrowing results from the earlier chapters. Unfortunately that is not the case. RLC circuits give rise to second-order linear differential equations. Solutions of these equations change shape drastically depending on the values of the circuit elements. Since electrical energy can be shuttled back and forth between an electrical field of a capacitor and a magnetic field of an inductor, circuit variables in a RLC circuit can begin to oscillate. In the first section of this chapter we discuss the general form of the solution of a second-order linear differential equation. In the subsequent sections these results are used to study the responses of RLC circuits.

10.1 Second-Order Linear Differential Equation

The general form of a second-order linear differential equation is

$$d^2\mathbf{x}/dt^2 + 2p \; d\mathbf{x}/dt + q^2 \; \mathbf{x}(t) = \mathbf{K}, \qquad (10.1.1)$$

where $\mathbf{K}$ is a constant and $\mathbf{x}(t)$ is an unknown function that satisfies (10.1.1). To find $\mathbf{x}(t)$ we assume that it is made up of two parts: a) the *transient solution*, and b) the *steady-state solution*.

Steady-state solution: We assume that in the steady-state the function $\mathbf{x}(t)$ has the same form as the input. In (10.1.1) the input K is a constant. Hence, in stady-state, the function $\mathbf{x}(t)$ becomes some unknown constant B. Since this steady-state solution must also satisfy (10.1.1), we substitute B for $\mathbf{x}(t)$ in (10.1.1) and get $q^2 \; B = \mathbf{K}$ or $B = \mathbf{K} / q^2$.

Transient solution: The transient solution of (10.1.1) is much more complex than that of (9.2.1). We assume that the transient solution is the result of nonzero initial value of $x(t)$ and any discontinuity in the input at start. It is called a transient solution because it decays with time. However, it can take many different forms which we shall have to investigate. To find the transient solution we solve

$$d^2x/dt^2 + 2p\ dx/dt + q^2\ x(t) = 0. \tag{10.1.2}$$

Since $\exp(st)$ worked well with (9.2.2) we shall try that on (10.1.2). Substituting $x(t) = \exp(st)$ in (10.1.2) we get

$$(s^2 + 2p\ s + q^2)\ \exp(st) = 0. \tag{10.1.3}$$

Since $\exp(st)$ is not identically zero for all values of t, in order to satisfy (10.1.3) we must select s such that

$$(s^2 + 2p\ s + q^2) = 0. \tag{10.1.4}$$

The left hand side (LHS) of (10.1.4) is a quadratic polynomial in s and we need to find its roots. The general expression for these roots are derived in the Appendix A. The roots are given by

$$s_1 = -p + (p^2 - q^2)^{1/2}, \tag{10.1.5}$$

and

$$s_2 = -p - (p^2 - q^2)^{1/2}, \tag{10.1.6}$$

and hence, the general form of the transient solution of (10.1.2) is

$$x(t) = A_1\ \exp(s_1 t) + A_2\ \exp(s_2 t), \tag{10.1.7}$$

where A_1 and A_2 are unknown constants. Thus we have a whole family of transient solutions parametrized by A_1 and A_2. The exact values of A_1 and A_2 are found utilizing known values of $x(t)$ and its derivatives.

The total solution of (10.1.1) becomes

$$x(t) = A_1\ \exp(s_1 t) + A_2\ \exp(s_2 t) + K\ /\ q^2. \tag{10.1.8}$$

10.2 Special Cases

The roots of (10.1.4) can be distinct and real, identical, or complex conjugates of each other (consult Appendix A for a short review of complex numbers and Euler identity). Hence, the transient solution can take three different shapes.

Distinct real roots: Assume $p^2 - q^2 > 0$. Then s_1 and s_2 are negative real numbers with s_2 being smaller than s_1. In this case both $\exp(s_1 t)$ and $\exp(s_2 t)$ are decaying exponentials. This case is usually called the *over-damped case*.

Complex conjugate roots: Assume $p^2 - q^2 < 0$. Then s_1 and s_2 are complex numbers and they are conjugates of each other. Let $\omega^2 = q^2 - p^2$. Then

$$A_1 \exp(s_1 t) + A_2 \exp(s_2 t)$$
$$= \exp(-pt) \, [A_1 \exp(j\omega t) + A_2 \exp(-j\omega t)] \qquad (10.2.1)$$

Once the values of A_1 and A_2 are known, the terms within the [ ] brackets reduce to some form of sinusoids. Since these sinusoids are multiplied by $\exp(-pt)$ we have sinusoids whose amplitudes change exponentially with time. If $p > 0$, then these become exponentially decaying sinusoids. This case is usually called the *under-damped case*.

Equal roots: Assume $p^2 - q^2 = 0$. In this case the transient solution does not have the form given by (10.1.7). We shall skip the details of the mathematical derivation in this case and state the form of the transient solution directly. The transient solution is given by

$$x(t) = A_1 \, t \exp(-pt) + A_2 \exp(-pt). \qquad (10.2.2)$$

This case is usually called the *critically-damped case*.

A parameter $\xi = p/q$, called the *damping-ratio* is often used to describe the state of damping in a second-order linear differential equation. From the three special cases discussed above, we see that $\xi = 1$ denotes critical damping. In a similar manner $\xi > 1$ and $\xi < 1$ denote over and under damping respectively.

In case of an under-damped system, q is called its natural frequency of oscillation. Note that the actual frequency of oscillation is ω where $\omega^2 = q^2 - p^2$. Hence, the actual frequency of oscillation ω is less than the natural frequency q.

10.3 Natural Response of a Series RLC

A series RLC circuit is shown in Figure 10.1. Since we are interested in its natural response, there are no sources present in the circuit. We assume that at $t = 0$, $v_C(0) = V_0$ and $i(0) = I_0$. Applying Kirchhoff's voltage law around the loop we get

$$L\ di/dt + R\ i(t) + v_C(0) + [\int_0^t i(x)dx]\ /\ C = 0. \tag{10.3.1}$$

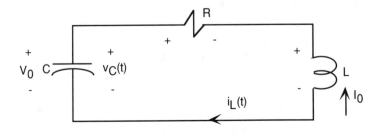

Figure 10.1. Series RLC circuit.

Differentiating both sides of (10.3.1) with respect to time we have

$$L\ d^2i/dt^2 + R\ di/dt + (1/C)\ i(t) = 0,$$

or

$$d^2i/dt^2 + (R/L)\ di/dt + (1/LC)\ i(t) = 0. \tag{10.3.2}$$

Here $p = R/2L$ and $q^2 = 1/LC$. Since under-damped is an interesting case we have not encountered in earlier chapters we assume that the circuit is under-damped i.e., $(R/2L)^2 < (1/LC)$.

Using (10.2.1) we get

$$i(t) = \exp(-pt)\ [A_1 \exp(j\omega t) + A_2 \exp(-j\omega t)] \tag{10.3.3}$$

Since we are considering circuits with both inductors and capacitors it is natural to assume that $L > 0$ and $C > 0$. therefore as long as $R > 0$, we see that $p = R/2L > 0$ and $\exp(-pt)$ is a decaying exponential. When $R = 0$, $p = 0$ and $i(t)$ does not decay with time.

To find the values of the unknown constants A_1 and A_2 we have to use values of $i(t)$ as well as its derivative. We know that $i(0) = I_0$ and applying this condition to (10.3.3) we get

$$I_0 = A_1 + A_2. \qquad (10.3.4)$$

By Kirchhoff's voltage law, the terminal voltage of the inductor at t=0 equals $R\ i(0) + v_C(0) = R\ I_0 + V_0$. However, this terminal voltage is also $L di/dt$ evaluated at t=0 where $i(t)$ is given by (10.3.3). Using this relation we get

$$R\ I_0 + V_0 = j\ \omega L\ (A_1 - A_2). \qquad (10.3.5)$$

Solving (10.3.4) and (10.3.5) for A_1 and A_2 we get $A_1 = [V_0 + (R + j\omega L)\ I_0]\ /\ j2\omega L$, $A_2 = A_1^*$ and

$$i(t) = \exp(-pt)\ 2\ \text{Re}[\ A_1\ \exp(j\omega t)\ ]. \qquad (10.3.6)$$

From (10.3.6) we conclude that $i(t)$ is a sinusoid with exponentially decaying amplitude. A simpler result can be obtained by assuming $I_0 = 0$. In this case $A_1 = V_0\ /\ j2\omega L$ and $A_1 \exp(j\omega t) = (V_0/2\omega L)\ [\sin(\omega t) - j\cos(\omega t)]$. Hence from (10.3.6) we obtain

$$i(t) = \exp(-pt)\ (V_0/\omega L)\ \sin(\omega t). \qquad (10.3.7)$$

The graph of $i(t)$ vs. t is shown in Figure 10.2. In case $p = 0$ we have

$$i(t) = (V_0/\omega L)\ \sin(\omega t), \qquad (10.3.8)$$

which is a sinusoid with constant amplitude of $V_0/\omega L$ shown in Figure 10.3. In a series RLC circuit, $p = 0$ implies $R = 0$, i.e., there is no resistance in the circuit. Consequently there is no dissipation of energy. The electrical energy simply shuttles back and forth between the electrical field of the capacitor and the magnetic field of the inductor. The circuit variables oscillate in a sinusoidal manner similar to the current given by (10.3.8).

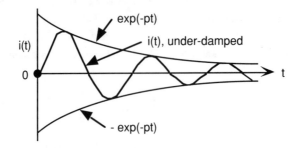

Figure 10.2. Under-damped natural response.

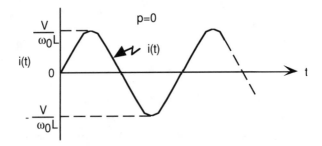

Figure 10.3. Natural response with zero damping.

10.4 Natural Response of a Parallel RLC

A parallel RLC circuit is shown in Figure 10.4. Since we are interested in its natural response, there are no sources present in the circuit. We assume that at t=0, $v(0) = V_0$ and $i_L(0) = I_0$. Applying Kirchhoff's current law at the top node we get

$$C \, dv/dt + v(t) / R + I_0 + [\int_0^t v(x)dx] / L = 0. \qquad (10.4.1)$$

Differentiating both sides of (10.4.1) with respect to time we have

$$\mathbf{C}\ d^2v/dt^2 + (1/\mathbf{R})\ dv/dt + (1/\mathbf{L})\ v(t) = 0,$$

or

$$d^2v/dt^2 + (1/\mathbf{RC})\ dv/dt + (1/\mathbf{LC})\ v(t) = 0, \qquad (10.4.2)$$

Here $p = 1/2\mathbf{RC}$ and $q^2 = 1/\mathbf{LC}$. Since critically-damped is an interesting case we have not encountered in earlier chapters we assume that the circuit is critically-damped i.e., $(1/2\mathbf{RC})^2 = (1/\mathbf{LC})$.

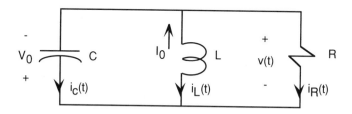

Figure 10.4. Parallel RLC circuit.

Using (10.2.2) we get

$$v(t) = A_1\ t\ \exp(-pt) + A_2\ \exp(-pt). \qquad (10.4.3)$$

To find the values of the unknown constants A_1 and A_2 we have to use values of $v(t)$ as well as its derivative. We know that $v(0) = \mathbf{V}_0$ and applying this condition to (10.4.3) we get

$$\mathbf{V}_0 = A_2. \qquad (10.4.4)$$

By Kirchhoff's current law, the terminal current of the capacitor at t=0 equals $v(0)/\mathbf{R} + i_L(0) = \mathbf{V}_0/\mathbf{R} + \mathbf{I}_0$. However, this terminal current is also $\mathbf{C}dv/dt$ evaluated at t=0 where $v(t)$ is given by (10.4.3). Using this relation we get

$$\mathbf{V}_0/\mathbf{R} + \mathbf{I}_0 = \mathbf{C}\ (A_1 - p\ A_2), \qquad (10.4.5)$$

and

$$v(t) = [\ V_0 + (3/2 \ V_0/RC + \ I_0/C) \ t] \ exp(-pt). \tag{10.4.6}$$

A simpler expression can be obtained by assuming $V_0 = 0$. In this case we have

$$v(t) = (I_0/C) \ t \ exp(-pt), \tag{10.4.7}$$

a graph of which is shown in Figure 10.5.

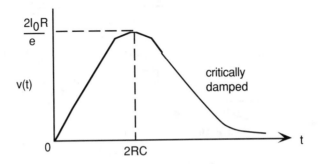

Figure 10.5. Critically damped response.

10.5 Step-Response of a Series RLC

A series RLC circuit with a constant source is shown in Figure 10.6. The switch S is closed at t=0 which applies a step voltage input to the circuit. Noting that $i(t) = C \ dv_C/dt$ and applying Kirchhoff's voltage law to the loop we get

$$LC \ d^2v_0/dt^2 + R \ dv_0/dt + v_0(t) = V_S, \tag{10.5.1}$$

where V_S is the output of the constant voltage source. Dividing both sides of (10.5.1) by LC we have

$$d^2v_0/dt^2 + (R/L)\, dv_0/dt + (1/LC)\, v_0(t) = V_S/LC, \qquad (10.5.2)$$

which is a special case of (10.1.1) with $p = (R/2L)$, $q^2 = (1/LC)$ and $K = V_S/LC$. Assuming that the circuit is under-damped, and using (10.1.8) and (10.2.1) we get

$$v_0(t) = A_1 \exp(s_1 t) + A_2 \exp(s_2 t) + V_S,$$

or

$$v_0(t) = \exp(-pt)\, [A_1 \exp(jwt) + A_2 \exp(-jwt)] + V_S. \qquad (10.5.3)$$

To simplify the process of finding A_1 and A_2 we assume that $v_C(0) = 0$ and the initial current in the inductor $I_0 = 0$. Since the current in the inductor can not change instantaneously and the inductor is in series with the capacitor $C\, dv_C/dt = 0$. Applying these two conditions to (10.5.3) we obtain

$$0 = A_1 + A_2 + V_S, \qquad (10.5.4)$$

and

$$0 = (-p)\,(A_1 + A_2) + j\omega\,(A_1 - A_2). \qquad (10.5.5)$$

Solving (10.5.4) and (10.5.5) for A_1 and A_2 and substituting in (10.5.3) we get

$$v_C(t) = V_S\, [\, 1 - (\exp(-pt)/\omega)\, (p \sin(\omega t) + \omega \cos(\omega t))\,]. \qquad (10.5.6)$$

If $R > 0$, then $p > 0$ and the transient part in (10.5.6.) is an exponentially decaying sinusoid. The steady-state value of $v_C(t)$ is V_S. $v_C(t)$ given by (10.5.6) is graphed in Figure 10.5.(b). Various important response parameters such as overshoot and rise-time are also shown in Figure 10.7.

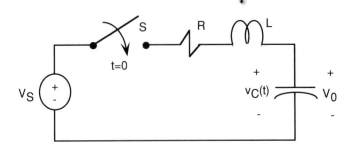

Figure 10.6. Series RLC circuit with a constant source.

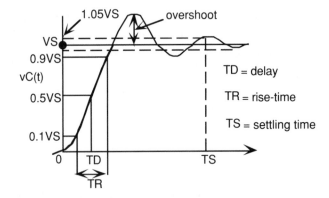

Figure 10.7. Under-damped step-response.

10.6 Step-Response of a Parallel RLC

A parallel RLC circuit with a constant source is shown in Figure 10.8. The switch S is closed at t=0 which applies a step voltage input to the circuit. Noting that $\mathbf{v}(t) = \mathbf{L}$ di_L/dt and applying Kirchhoff's current law to the top node we get

$$\mathbf{LC}\ d^2i_L/dt^2 + \mathbf{L/R}\ di_L/dt + i_L(t) = \mathbf{I_S}, \qquad (10.6.1)$$

where I_S is the output of the constant current source. Dividing both sides of (10.5.1) by LC we have

$$d^2i_L/dt^2 + (1/RC) \, di_L/dt + (1/LC) \, i_L(t) = I_S/LC, \qquad (10.6.2)$$

which is a special case of (10.1.1) with $p = (1/2RC)$, $q^2 = (1/LC)$ and $K = I_S/LC$. Assuming that the circuit is under-damped, and using (10.1.8) and (10.2.1) we get

$$i_L(t) = A_1 \, \exp(s_1 t) + A_2 \, \exp(s_2 t) + I_S,$$

or

$$i_L(t) = \exp(-pt) \, [A_1 \, \exp(j\omega t) + A_2 \, \exp(-j\omega t)] + I_S. \qquad (10.6.3)$$

To simplify the process of finding A_1 and A_2 we assume that $v_C(0) = 0$ and the initial current in the inductor $I_0 = 0$. Since the voltage across the capacitor can not change instantaneously and the capacitor is in parallel with the inductor $L \, di_L/dt = 0$. Applying these two conditions to (10.6.3) we obtain

$$0 = A_1 + A_2 + V_S, \qquad (10.6.4)$$

and

$$(-p) \, (A_1 + A_2) + j\omega \, (A_1 - A_2) = 0. \qquad (10.6.5)$$

Solving (10.6.4) and (10.6.5) for A_1 and A_2 and substituting in (10.6.3) we get

$$i_L(t) = I_S \, [\, 1 - (\exp(-pt)/\omega) \, (p \sin(\omega t) + \omega \cos(wt)) \,]. \qquad (10.6.6)$$

If $R > 0$, then $p > 0$ and the transient part in (10.6.6.) is an exponentially decaying sinusoid. The steady-state value of $i_L(t)$ is I_S. $i_L(t)$ given by (10.6.6) has exactly the same shape as $v_C(t)$ given by (10.5.6). Hence, Figure 10.7 with obvious modifications also serves to illustrate $i_L(t)$.

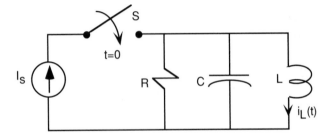

Figure 10.8. Parallel RLC circuit with a constant source.

10.7 SPICE Simulation of RLC Circuits

SPICE can be used to simulate the natural and the step-responses of a whole family of RLC circuits. The effect of the damping-ratio on the shape of the response becomes clear at a glance from the graphical output of such a simulation. Figure 10.9 shows the circuit file of a family of RLC circuits simulated simultaneously. The natural responses are plotted on the same pair of axes in Figure 10.10. The L and the C values are held constant and the R value is changed to produce the three types of responses. Figure 10.11 shows the step-responses of the same circuits for varying damping-ratios.

```
rclnat.cir natural response of RCL circuits

* under-damped section
r1   0    1     0.3K
c1   1    2     0.001M      ic = 4V
l1   2    3     250M        ic = 0A
vd1  0    3     dc    0V

* critically-damped section
r2   0    4     1K
c2   4    5     0.001M      ic = 4V
l2   5    6     250M        ic = 0A
vd2  0    6     dc    0V
```

Figure 10.9. Circuit file of a family of RLC circuits simulated simultaneously.

```
* over-damped section
r3    0      7      3K
c3    7      8      0.001M        ic = 4V
13    8      9      250M          ic = 0A
vd3   0      9      dc     0V

.tran   1M   10M   uic
.probe  i(vd1)  i(vd2)  i(vd3)
.options nopage
.end
```

Figure 10.9. Circuit file of a family of RLC circuits simulated simultaneously (cont).

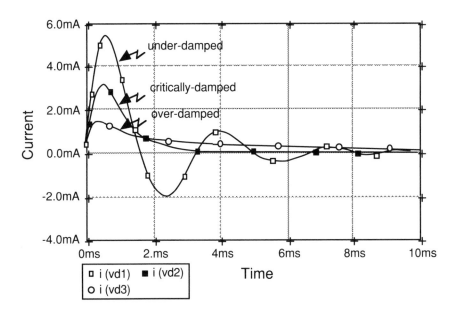

Figure 10.10. Natural response of RLC circuits.

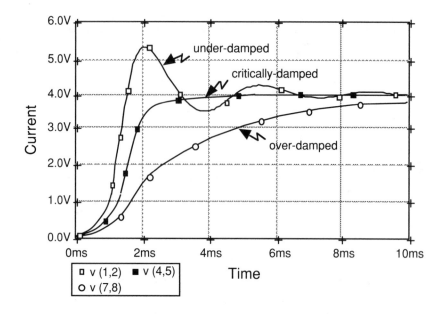

Figure 10.11. Step-response of RLC circuits.

10.8 Examples

Example 10.8.1

Analyze the RLC circuit shown in Figure 10.12. We note that this is not a standard series or parallel RLC circuit. However, in some cases such as this one, it is possible to obtain a standard series or parallel form after the application of Thevenin's or Norton's theorem.

The Thevenin equivalent circuit is shown in Figure 10.13 which clearly is of the standard series RLC form. Hence, the natural response or the step-response can be obtained by using the methods described in the previous sections.

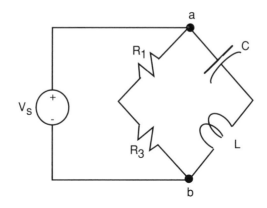

Figure   10.12.

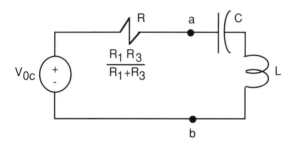

Figure   10.13.

Example 10.8.2

   Nonstandard RLC circuits, such as the one shown in Figure 10.14, can be analyzed by the usual methods of circuit analysis discussed so far.

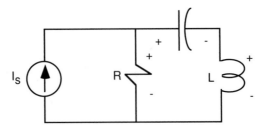

Figure   10.14.

Let us assume that $v_C(0) = 0$ and $i_L(0) = 0$. Then at t=0 all the source current I_S will go through the resistor R and $v_L(0) = I_S R$. In general

$$L \ di/dt + [\int_0^t i(x)dx \ ] \ / \ C + R \ i(t) = R \ I_S.$$

or

$$L \ d^2i/dt^2 + R \ di/dt + (1/C) \ i(t) = 0. \tag{10.8.1}$$

Alternately, the circuit can be converted to a standard series RLC form by applying source transformation.

Example 10.8.3

In Figure 10.15 we assume that $v_C(0) = - V_0$ and $i_L(0) = 0$. Applying Kirchhoff's voltage law we have

$$L \ di/dt - V_0 + [\int_0^t i(x)dx \ ] \ / \ C = 0,$$

or

$$d^2i/dt^2 + (1/LC) \ i(t) = 0. \tag{10.8.2}$$

In this case $p = 0$ and the circuit must be under-damped. Setting $p = 0$ in (10.3.3) we get

$$i(t) = [A_1 \exp(j\omega t) + A_2 \exp(-j\omega t)] \tag{10.8.3}$$

Since $i_L(0) = 0$ is given and $L\, di/dt\,|_{t=0} = v_L(0) = - v_C(0) = V_0$, the two equations for A_1 and A_2 become

$$0 = A_1 + A_2, \tag{10.8.4}$$

and

$$V_0/L = j\omega\,(A_1 - A_2). \tag{10.8.5}$$

Solving (10.8.4) and (10.8.5) for A_1 and A_2 and substituting in (10.8.3) we obtain

$$i(t) = (V_0/\omega L)\,\sin(\omega t)]. \tag{10.8.6}$$

In the absence of any resistor in the circuit, no part of the initial energy in the capacitor is ever dissipated. The energy shuttles back and forth between the electrical field and the magnetic field. Hence, the circuit variables oscillate like sinusoids.

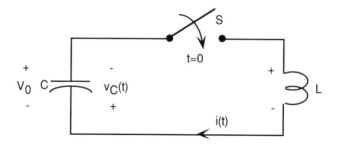

Figure   10.15.

10.9 Summary

•   The form of the solution of a second-order linear differential equation depends on the parameter values. Three types of solution are possible: 1) over-damped, 2) critically-damped, and 3)under-damped.

• The type of damping is decided by the value of the damping-ratio ξ. In the under-damped case the frequency of oscillation is given by ω.

• In case of an under-damped circuit, a circuit variable may overshoot its steady-state value. Several characteristics of such a response such as over-shoot, rise-time, and delay-time are shown in Figure 10.5.(b).

• In case of a critically-damped circuit, the response is given by $x(t) = A_1\, t \exp(-pt) + A_2 \exp(-pt)$. Figure 10.4.(b) shows a typical case of a critically-damped response.

• If the RLC circuit is not in a series or parallel form, then it can often be transformed into one by means of Thevenin or Norton theorem. Otherwise, conventional circuit analysis techniques such as node equations or voltage/current divisions can be directly used to obtain the required differential equations.

10.10   Problems

10.1.   In a series RLC circuit with a constant voltage source as input, the sum of the terminal voltages of the elements should equal the source output according to Kirchhoff's voltage law. In case of an under-damped circuit, the voltage across the capacitor overshoots the source output (see Figure 10.5.(b)). Explain how this is possible?

10.2.   Using (10.5.6) find an expression for the current in an under-damped series RLC circuit. From this expression derive the steady-state value of the current. Justify this steady-state value from physical considerations.

10.3.   Using the expression for current derived in Problem 10.10.2 find the terminal voltages of the resistor and the inductor respectively. Verify that Kirchhoff's voltage law is satisfied.

10.4.   From (10.4.7) find expressions for the terminal currents of the capacitor and the resistor respectively. From these two expressions directly compute the terminal current of the inductor. Verify that the terminal current and the terminal voltage of the inductor satisfy $Ldi/dt = v(t)$.

10.5.   Using (10.5.6) find an expression for the overshoot and the time t for its
        occurrence.

10.6.   Assume that the RLC circuit is very much under-damped. Using (10.5.6) find an
        expression for the time-delay which equals the time taken by $v_C(t)$ to reach 50%
        of its steady-state value. (Hint: assume $\sin(\omega t) = 0$ and $\cos(\omega t) = 1$).

10.7.   The voltage in a critically-damped parallel RLC circuit is given by (10.4.7). Find
        an expression for the maximum value of the voltage and the time t of its
        occurrence.

10.8.   Find an expression for $v(t)$ in the circuit shown in Figure 10.16. Assume that
        the switch S is closed at t=0 and $v_C(0) = i_L(0) = 0$. (Hint: Use node equation at
        node 1. Note $i(t) = I_S - (v(t)/R) - Cdv/dt$.)

10.9.   Find an expression for $v(t)$ in the circuit shown in Figure 10.17. Assume that
        the switch S is closed at t=0 and $v_C(0) = i_L(0) = 0$.

10.10. Find an expression for $v(t)$ in the circuit shown in Figure 10.18. Assume that
        the switch S is closed at t=0 and $v_C(0) = i_L(0) = 0$. (Hint: Use Thevenin's or
        Norton's equivalent circuit.)

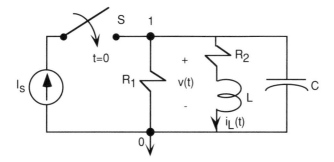

Figure   10.16.

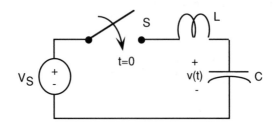

Figure 10.17.

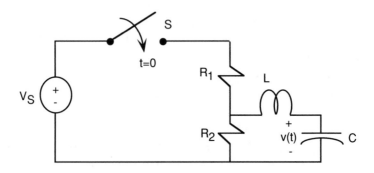

Figure 10.18.

11

SINUSOIDAL ANALYSIS

Analysis of linear circuits containing resistors, capacitors, and inductors and energized by sinusoidal input sources is important at least for two reasons: 1) electrical energy is commonly distributed in the sinusoidal form, and 2) responses of circuits to nonsinusoidal sources can be predicted on the basis of their sinusoidal responses. Various techniques of circuit design are also based on the sinusoidal analysis of circuits. This chapter presents the basic approach to the sinusoidal analysis of circuits.

11.1 Sinusoidal Sources

Sinusoidal voltage and current sources were introduced in Section 7.6. A sinusoidal source is given by

$$\mathbf{v}(t) = \mathbf{V} \, \sin(\omega t + \theta), \qquad (11.1.1)$$

or

$$\mathbf{i}(t) = \mathbf{I} \, \cos(\omega t + \theta). \qquad (11.1.2)$$

Either the sine or the cosine function can be used to define a sinusoid. The *amplitudes* of the sinusoids are $\mathbf{V}$ and $\mathbf{I}$ respectively with a common *angular frequency* of ω *radians/sec* and *phase* θ *radians*. Figure 11.1 shows a sinusoid given by (11.1.1).

A sinusoid is often specified by its frequency $\mathbf{f}$ in cycles/sec or *Hertz* where $\omega = 2\pi$ $\mathbf{f}$ and the time period is $\mathbf{T} = 1/\mathbf{f}$. A phase angle θ is a measure of the displacement of the sinusoid along the time axis from the origin. A phase angle of θ radians corresponds to a displacement in time of $t_D = -\theta/\omega$ seconds. A positive value of θ gives a time advance or time lead ($t_D < 0$) and θ is called a *phase lead*. A negative value of θ gives a time delay or time lag ($t_D > 0$) and θ is called a *phase lag*. Figure 11.2 shows sinusoids which are delayed or advanced with respect to the standard sinusoid sin(ωt). The value of θ may be given either in *degrees* or in *radians*. However, it must be converted to radians before it is actually added to ωt.

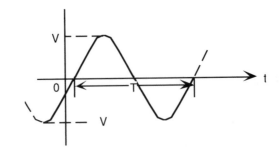

Figure 11.1. V sin(ωt + θ) vs. t.

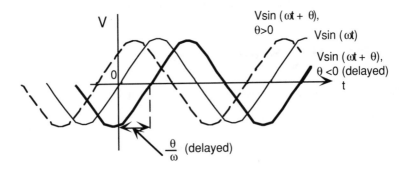

Figure 11.2. Delayed and advanced sinusoids.

11.2 Sinusoidal Response

As an example of a sinusoidal analysis of a circuit, consider the circuit shown in Figure 11.3. The first-order linear differential equation satisfied by the voltage v(t) across the capacitor is

$$dv(t)/dt + (1/\mathbf{RC})\ v(t) = \mathbf{V/RC}\ \sin(\omega t) \qquad (11.2.1)$$

From Section 9.2 we know that (11.2.1) has a transient solution that decays exponentially with time. This transient response is close to zero within a few time-constants. By the *sinusoidal response* we mean the steady-state solution of (11.2.1). This steady-state solution is different from what we had in Section 9.2 because the input is no longer a constant but a sinusoid.

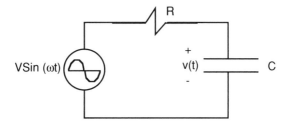

Figure 11.3.

Following the same line of reasoning as in Section 9.2 we assume that the steady-state response is of the form $v(t) = A\ \sin(\omega t + \theta)$. Substituting this solution in (11.2.1) and using trigonometric identities we get

$$A^2 = \mathbf{V}^2\ /\ (\ 1 + \omega^2\ \mathbf{R^2C^2}\ ) \qquad (11.2.2)$$

and

$$\theta = -\ \arctan(\ \omega\ \mathbf{RC}\ ). \qquad (11.2.3)$$

Note that the amplitude and the phase of the sinusoidal response of a circuit depend on the amplitude and the angular frequency of the source output as well as the parameters of the circuit. Once the amplitude and the phase of the sinusoidal response are known, the response is known for all time since it is a sinusoid of angular frequency ω.

11.3 Phasors

Sinusoidal analysis of a circuit amounts to the determination of the amplitude and the phase of the circuit response. We make use of trigonometric identities to derive expressions such as (11.2.2) and (11.2.3). Also some differentiation is necessary which transforms sine functions into cosine functions and vice versa (don't forget the negative sign). A much more straight forward and completely algebraic approach to sinusoidal analysis is possible if we make use of complex numbers (see Appendix A).

The Euler identity states

$$\exp(j\theta) = \cos(\theta) + j\,\sin(\theta). \tag{11.3.1}$$

Hence, a sinusoid such as

$$v(t) = A\,\sin(\omega t + \theta) \tag{11.3.2}$$

can be written as

$$\begin{aligned} \mathbf{v}(t) &= \mathrm{Im}\,[\ A\,\exp(\ j\,(\omega t + \theta)\ )\ ] \\ &= \mathrm{Im}\,[\ A\,\exp(j\theta)\,\exp(\ j\omega t\ )\ ], \end{aligned} \tag{11.3.3}$$

where Im denotes the imaginary part of a complex expression. The part A exp(jθ) is a complex number and is called a stationary phasor or *phasor* for short. exp(jωt) is called a rotating phasor. Figure 11.4 shows these two phasors on the complex plane. We can define a phasor transformation 'Ph' from sinusoids to phasors as follows:

$$\mathrm{Ph}[\ \mathbf{v}(t)\ ] = A\,\exp(j\theta) \tag{11.3.4}$$

where v(t) is a sinusoid given by (11.3.2). The inverse transform requires the knowledge of whether the sinusoids are sine or cosine functions. Assuming that the sinusoids used are all sine functions we get

$$\mathrm{Ph}^{-1}\,[A\,\exp(\ j\theta)] = \mathrm{Im}\,[\ A\,\exp(\ j\theta)\,\exp(\ j\omega t\ )\ ]. \tag{11.3.5}$$

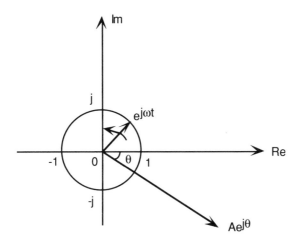

Figure 11.4. Phasors in the complex plane.

11.4 R, C, and L in the Phasor Domain

In analyzing circuits, the relationships between terminal voltages and terminal currents of circuit elements have been found to be most useful. Let us determine what these relationships are for resistors, capacitors and inductors once sinusoids are replaced by phasors.

For a resistor, by Ohm's law $v(t) = \mathbf{R}\ i(t)$ for all time t. Hence, given that $i(t) = \mathbf{I}$ $\sin(\omega t + \theta)$, $v(t) = \mathbf{R}\ i(t) = \mathbf{R}\mathbf{I}\ \sin(\omega t + \theta)$. It follows that Ph[$v(t)$] = $\mathbf{R}$ Ph[$i(t)$] and hence, that Ohm's law remains unchanged except that voltages and currents are represented by complex numbers called phasors.

In case of a capacitor $i(t) = \mathbf{C}\ dv(t)/dt$. Given that $v(t) = \mathbf{V}\ \sin(\omega t + \theta)$, we have $i(t)$ $= \omega\mathbf{C}\ \mathbf{V}\ \cos(\omega t + \theta) = \omega\mathbf{C}\ \mathbf{V}\ \sin(\omega t + \theta + \pi/2)$. Hence, Ph[$i(t)$] = $\omega\mathbf{C}\ \mathbf{V}$ Ph[$\sin(\omega t$ $+ \theta + \pi/2)$] = $\omega\mathbf{C}\ \exp(j\pi/2)$ Ph[$\mathbf{V}\ \sin(\omega t + \theta)$] = $j\omega\mathbf{C}$ Ph[$v(t)$]. Note that now we have an algebraic relationship between the current phasor Ph[$i(t)$] and the voltage phasor Ph[$v(t)$]. Let $\mathbf{X}$ denote the phasor Ph[$x(t)$]. Then this algebraic relationship can be written explicitly as

$$I = j\omega C \ V \tag{11.4.1}$$

Equation (11.4.1) is sort of an Ohm's law for a capacitor in the phasor domain. It says that in terms of phasors, a capacitor has a *reactance* given by $V/I = 1/j\omega C = -j$ $(1/\omega C)$. In a similar manner we can show that in the phasor domain an inductor has a *reactance* given by $V/I = j\omega L$, i.e.

$$V = j\omega L \ I \tag{11.4.2}$$

A reactance is an imaginary number and can be shown on the complex plane. Figure 11.5 shows the resistance and the two reactances of **R**, **C**, and **L** on the complex plane. Similar purely algebraic relationships can be used to describe the terminal relationships of controlled sources in the phasor domain.

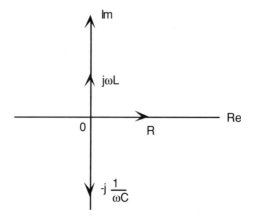

Figure 11.5. Resistance and reactance in the complex plane.

11.5 Circuit Laws

We observe from the previous section that, in the phasor domain, the terminal voltage/current relationships of resistors, capacitors, and inductors are all given by algebraic equations of the same form $V/I = $ constant where the constants are either real or imaginary numbers. Now we investigate the transformations of Kirchhoff's laws in the phasor domain.

Let $v_k(t)$ denote the sinusoidal voltage drop across the kth branch included in a closed loop. Then $v_k(t) = \text{Im} [\mathbf{V}_k \exp(j\omega t)]$ where $\mathbf{V}_k$ is the corresponding phasor. Sum of such voltages around a loop can be written as $v_1(t) + ... + v_n(t) = \text{Im} [(\mathbf{V}_1 + ... + \mathbf{V}_n) \exp(j\omega t)]$. By Kirchhoff's voltage law, the left hand side of this sum is zero. Hence, on the right hand side we must have $\mathbf{V}_1 + ... + \mathbf{V}_n = 0$, i.e. the voltage phasors around a closed loop satisfy Kirchhoff's voltage law.

Let $i_k(t)$ denote the sinusoidal current through the kth branch joined at a node. Then $i_k(t) = \text{Im} [\mathbf{I}_k \exp(j\omega t)]$ where $\mathbf{I}_k$ is the corresponding phasor. Sum of such currents at a node can be written as $i_1(t) + ... + i_n(t) = \text{Im} [(\mathbf{I}_1 + ... + \mathbf{I}_n) \exp(jwt)]$. By Kirchhoff's current law, the left hand side of this sum is zero. Hence, on the right hand side we must have $\mathbf{I}_1 + ... + \mathbf{I}_n = 0$, i.e. the current phasors at a node satisfy Kirchhoff's current law.

Now looking back at the material in Chapters 2, 4, 5, and 6 we note that all of it was established in terms of Kirchhoff's laws, linearity and simple algebraic equations (Ohm's law) describing the terminal relationships of circuit elements. Hence, all this material remains valid in the phasor domain as long as we make use of complex numbers rather than real numbers.

11.6 Impedance and Admittance

In Section 11.4 we found that a relation $\mathbf{V}/\mathbf{I} = \text{constant}$, similar to the one given by Ohm's law for resistors, is valid in the phasor domain for R, C, and L. In general this ratio is a complex number and is called an *impedance*.

$$\mathbf{V}/\mathbf{I} = \mathbf{Z}(j\omega)$$
$$= \mathbf{R}(\omega) + j\, \mathbf{X}(\omega) \qquad (11.6.1)$$

The expression $\mathbf{R}(\omega)$ is called the *resistive* part of the impedance. It is not nescessarily made out of only resistor values. The expression $\mathbf{X}(\omega)$ is called the *reactive part* of the impedence. The reciprocal of an impedance is called an *admittance*.

$$\mathbf{I}/\mathbf{V} = \mathbf{Y}(j\omega)$$
$$= \mathbf{G}(\omega) + j\, \mathbf{B}(\omega) \qquad (11.6.2)$$

$G(\omega)$ and $B(\omega)$ are respectively the *conductive* and the *susceptive* parts of the admittance. Since the material of Chapter 2 is valid in the phasor domain, the formulas for the series and the parallel combination of impedances have the same form as those in (2.3.5) and (2.4.5).

Series: $\mathbf{Z}_{eq}(j\omega) = \mathbf{Z}_1(j\omega) + \mathbf{Z}_2(j\omega)$ (11.6.3)

Parallel: $\mathbf{Z}_{eq}(j\omega) = \mathbf{Z}_1(j\omega)\,\mathbf{Z}_2(j\omega)\,/\,(\,\mathbf{Z}_1(j\omega) + \mathbf{Z}_2(j\omega))$ (11.6.4)

Example 11.6.1

Derive an expression for the impedance of the series connection of a resistor and an inductor shown in Figure 11.6.

In the phasor domain $\mathbf{V}_R\,/\,\mathbf{I}_R = R$ and $\mathbf{V}_L\,/\,\mathbf{I}_L = j\omega\,L$. Furthermore $\mathbf{I}_S = \mathbf{I}_R = \mathbf{I}_L$ in a series connection and $\mathbf{V}_S = \mathbf{V}_R + \mathbf{V}_L$. Using these relations we have $\mathbf{Z}(j\omega) = \mathbf{V}_S\,/\,\mathbf{I}_S = R + j\omega\,L$. The same expression can also be derived using (11.6.3).

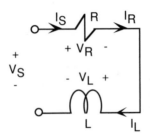

Figure 11.6. R in series with L.

Example 11.6.2

Derive an expression for the impedance of the parallel connection of a resistor and a capacitor shown in Figure 11.7.

From the circuit diagram $\mathbf{I}_C = j\omega\,C\,\mathbf{V}_S$, $\mathbf{I}_R = \mathbf{V}_S\,/\,R$, and $\mathbf{I}_S = \mathbf{I}_C + \mathbf{I}_S$. Hence, $\mathbf{Z}(j\omega) = \mathbf{V}_S\,/\,\mathbf{I}_S = R(\omega) + j\,X(\omega)$ where $R(\omega) = R\,/\,(1 + \omega^2 R^2 C^2)$, and $X(\omega) = -\,\omega R^2 C\,/\,(1 + \omega^2 R^2 C^2)$. The same expression can also be derived using (11.6.4).

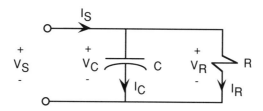

Figure 11.7. R in parallel with C.

Example 11.6.3

Derive an expression for the impedance of the parallel connection of an inductor and a capacitor in series with a resistor as shown in Figure 11.8.

From the circuit diagram $\mathbf{I_C} = j\omega\,C\,\mathbf{V}$, and $\mathbf{I_L} = \mathbf{V}\,/\,j\omega\,L$. By Kirchhoff's current law $\mathbf{I_R} = \mathbf{I_C} + \mathbf{I_L} = \mathbf{V}\,(1 - \omega^2 CL)\,/\,j\omega L$, and by Ohm's law $\mathbf{V_R} = \mathbf{R}\,\mathbf{I_R}$. The total voltage drop is $\mathbf{V_R} + \mathbf{V} = \mathbf{V}\,[j\omega L + \mathbf{R}\,(1 - \omega^2 CL)]\,/\,j\omega L$ and $\mathbf{Z}(j\omega) = (\mathbf{V_R} + \mathbf{V})\,/\,\mathbf{I_R} = \mathbf{R} + j[\omega L\,/\,(1 - \omega^2 CL)]$. The same expression can also be derived using (11.6.3) and (11.6.4).

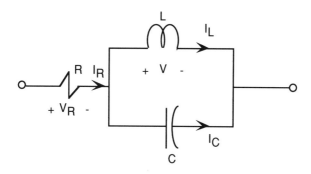

Figure 11.8. R in series with L and C in parallel.

11.7 Validation of Expressions

The examples above show that both the resistive part $R(\omega)$ and the reactive part $X(\omega)$ of an impedance are in general functions of the frequency ω of the sinusoid. This observation can be fruitfully used in testing the validity of expressions derived by algebra.

Following our discussion of validation of expressions in Chapter 2, we attempt to simplify the circuit being analyzed. Because of the sinusoidal input an additional technique for simplifying a circuit is to consider the case $\omega = 0$. At $\omega = 0$ the reactance of an inductor becomes zero. The voltage drop across the inductor becomes zero and hence, it can be considered as a short circuit.

Similarly at $\omega = 0$ the reactance of a capacitor becomes infinite. The current through the capacitor becomes zero making the capacitor into an open circuit. So at $\omega = 0$, the circuit becomes purely resistive and much easier to analyze. An alternate form of simplification results if we consider ω to be very high, approaching infinity. Then the reactance of an inductor approaches infinity and that of a capacitor becomes zero with obvious circuit simplifications.

Since the reactive part $X(\omega)$ of an impedance is a function of ω, it can become zero or infinite at frequencies dependent on the circuit parameters. This feature can also be used as a tool for the validation of expressions.

In Example 11.6.3, for $\omega^2 < 1/CL$ the reactance is of the form jA which is the form of an inductive reactance. For $\omega^2 > 1/CL$ it has the form $-jA$ which is the form of a capacitive reactance. For $\omega^2 = 1/CL$ the reactance is infinite. Hence, as the frequency ω of the input sinusoid increases from zero, the impedance $Z(j\omega)$ of the circuit changes from $R + j\omega L$ to $R - j\,1/\omega C$ (approx.).

For very low and very high values of ω, $Z(j\omega)$ is approximately R. The phase angle of $Z(j\omega)$ goes through a sharp change, in the vicinity of $\omega^2 = 1/CL$, from a positive to a negtive value. The phase angle of the terminal current of R changes from zero to a low negative value as ω approaches $1/CL$ from below. As ω crosses $1/CL$ and increases further, the phase angle of the current decreases from a high positive value back to zero again. These conditions can be verified in a SPICE simulation of the circuit.

11.8 SPICE Simulation of Sinusoidal Input

SPICE can be used to simulate the sinusoidal response of a circuit to a whole range of frequencies. The control line for sinusoidal input simulation is:

$$\textbf{.ac} \qquad \text{lin} \qquad \text{np} \qquad \text{fstart} \qquad \text{fstop}$$

The parameter **lin** indicates that the sinusoidal frequency is to be varied linearly over a range of frequencies given by **fstart** and **fstop.np** denotes the number of frequency points in the given range at which the response value is to be computed. For simulation at a single frequency we set **np** = 1 and **fstart** = **fstop** = frequency of input sinusoid in Hertz (**f**).

The output voltage and/or current phasors can be expressed either in polar or in rectangular form (see Appendix A.) The corresponding output lines in SPICE are:

$$\textbf{.print} \qquad \text{ac} \qquad \text{vm(n1,n2)} \quad \text{vp(n1,n2)}$$
$$\textbf{.print} \qquad \text{ac} \qquad \text{vr(n1,n2)} \quad \text{vi(n1,n2)}$$

The prefix **v** obviously indicates a voltage variable. For a current variable the prefix is **i**. Also **m** and **p** indicate the magnitude and the phase respectively and hence, the first output line prints the phasor voltage value between nodes n1 and n2 in *polar* form. In the second output line **r** and **i** indicate the real and the imaginary parts respectively and hence, this line prints the phasor voltage value between nodes n1 and n2 in *rectangular* form. **.print** can be augmented by a **.plot** command also. For better graphical output we shall use the **.probe**™ graphics command of our SPICE software.

Example 11.8.1

This is the simple circuit discussed as Example 11.6.3 and shown in Figure 11.8. The circuit file is shown in Figure 11.9. The input sinusoid frequency ω is varied over a range that covers the value of $[1/\textbf{CL}]^{1/2}$. The magnitude and the phase of the current phasor in the circuit are plotted in Figure 11.10. Since the input voltage is of amplitude one and phase zero, the current phasor is proportional to $1 / \textbf{Z}(j\omega)$. At $\omega^2 = 1/\textbf{CL}$, the

magnitude of the current phasor goes to zero. Its phase angle goes through a rapid change
from a negative to a positive value in the vicinity of $\omega^2 = 1/CL$. These observations are
in accordance with our analysis of the variation of $Z(j\omega)$ with ω given in Section 11.7.

```
sinul.cir   first sinusoidal circuit
* input sinusoidal source
* name   nodes   ac    amplitude   phase
  vs      1  0   ac        1         0

r1   1     2     5K
c1   3     0     1M
l1   3     0     1M
vd   2     3     dc      0V

.ac    lin    50    158    160
.print ac ip(vd)
.probe
.options   nopage
.end
```

Figure 11.9.   Circuit file.

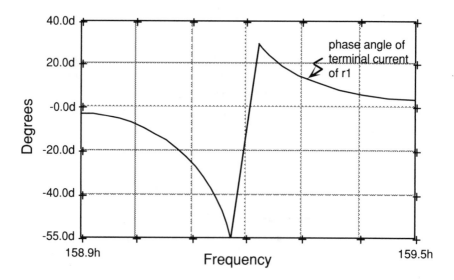

Figure 11.10. Magnitude and phase of the current phasor

Example 11.8.2

For our second example of SPICE simulation we consider the circuit shown in Figure 11.11. The input voltage is given in phasor form $\mathbf{V} \underline{/0°}$. Using the series and the parallel combination of impedances (11.6.3) and (11.6.4) we obtain the total impedance $\mathbf{Z}(j\omega) = \mathbf{R}_1 + [\mathbf{R}_2 (1 - \omega^2 \mathbf{LC})] / [(1 - \omega^2 \mathbf{LC}) + j\omega \mathbf{R}_2 \mathbf{C}]$. Clearly as ω goes to zero or infinity, $\mathbf{Z}(j\omega)$ approaches $\mathbf{R}_1 + \mathbf{R}_2$. and at $\omega^2 = 1/\mathbf{LC}$, $\mathbf{Z}(j\omega) = \mathbf{R}_1$.

At low values of ω, the capacitor has a very high impedance and the circuit is approximately a series combination of $\mathbf{R}_1$ and $\mathbf{R}_2$. Similarly, at very high values of ω the inductor has a very high impedance and the circuit again can be approximated as a series combination of $\mathbf{R}_1$ and $\mathbf{R}_2$.

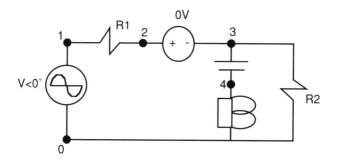

Figure 11.11. Input voltage given in phasor form.

Figure 11.12 shows a circuit file of the same structure as that shown in Figure 11.11. The magnitude and phase of $1/\mathbf{I}(j\omega)$ is plotted in Figure 11.13 (a and b) and compares well with our analysis.

```
sinu2.cir  second sinusoidal circuit
* input sinusoidal source
* name      nodes     ac   amplitude      phase
    vs        1  0     ac      1             0

r1   1     2     5
vd   2     3     dc    0V
c1   3     4     1M
l1   4     0     1M
r2   3     0     5

.ac   lin   500   1   10000
.print ac ip(vd)
.probe
.options   nopage
.end
```

Figure 11.12. Circuit file.

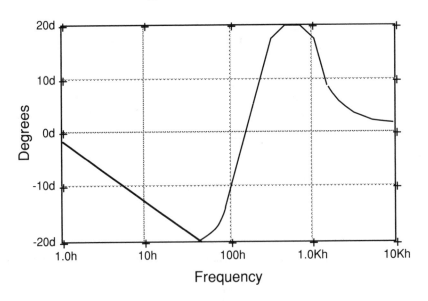

Figure 11.13 (a). Phase of $1/I(j\omega)$

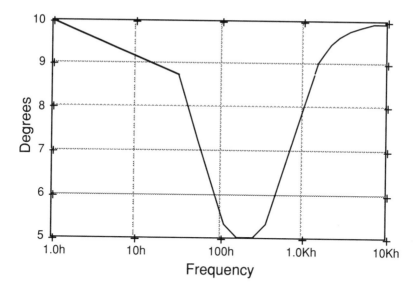

Figure 11.13 (b). Magnitude of $1/I(j\omega)$

11.9 Examples of Circuit Analysis

In this section we shall analyze circuits with sinusoidal sources in terms of the abstract impedance function $Z(j\omega)$. This approach clearly points out that the general techniques are still valid without unnecessary algebraic details clouding up the picture.

Example 11.9.1 Voltage Division.

In the circuit shown in Figure 11.13, the current phasor is $I(j\omega) = V(j\omega) / [Z_1(j\omega) + Z_2(j\omega)]$ and the voltage $V_2(j\omega) = V(j\omega) Z_2(j\omega) / [Z_1(j\omega) + Z_2(j\omega)]$.

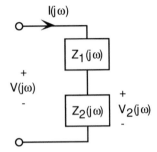

Figure 11.13. Impedances in series.

Example 11.9.2 Current division.

In the circuit shown Figure 11.14, the voltage phasor is $\mathbf{V}(j\omega) = \mathbf{I}(j\omega)\, \mathbf{Z}_1(j\omega)$ $\mathbf{Z}_2(j\omega) / [\mathbf{Z}_1(j\omega) + \mathbf{Z}_2(j\omega)]$ and the current phasor $\mathbf{I}_2(j\omega) = \mathbf{V}(j\omega) / \mathbf{Z}_2(j\omega) = \mathbf{I}(j\omega)\, \mathbf{Z}_1(j\omega)$ $/ [\mathbf{Z}_1(j\omega) + \mathbf{Z}_2(j\omega)]$.

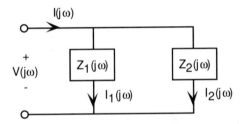

Figure 11.14. Impedances in parallel.

Example 11.9.3 Node equations.

All the techniques of writing node equations discussed before are still valid when the resistance **R** is replaced by the more general expression of an impedance $\mathbf{Z}(j\omega)$. Consider the circuit shown in Figure 11.15. The corresponding node equations are:

$$(\mathbf{V}_1 - \mathbf{V}) / \mathbf{Z}_1 + \mathbf{V}_1/\mathbf{Z}_2 - k\mathbf{I}_3 = 0,$$
$$k\mathbf{I}_3 + \mathbf{V}_2/\mathbf{Z}_3 - \mathbf{I} = 0,$$
$$\mathbf{I}_3 = \mathbf{V}_2/\mathbf{Z}_3.$$

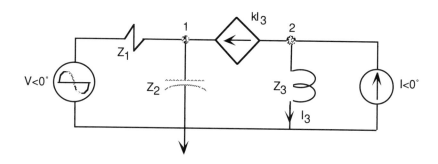

Figure 11.15. Example for node equations.

Example 11.9.4 Thevenin, Norton equivalent circuit.

In the circuit shown in Figure 11.16, under short-circuit conditions, the voltage phasor $V_2(j\omega) = 0$. Hence, the short-circuit current phasor becomes $I_{SC}(j\omega) = -k_1 I_1(j\omega)$ $= -k_1 V(j\omega) / Z_1(j\omega)$. The open-circuit voltage phasor satisfies the equation $V_{OC}(j\omega) = -k_1 I_1(j\omega) Z_2(j\omega) = -k_1 Z_2(j\omega) [V(j\omega) -k_2 V_{OC}(j\omega)] / Z_1(j\omega)$. Solving for $V_{OC}(j\omega)$ we have $V_{OC}(j\omega) = -k_1 Z_2(j\omega) V(j\omega) / [Z_1(j\omega) - k_1 k_2 Z_2(j\omega)]$. The equivalent impedance is $Z_{EQ}(j\omega) = V_{OC}(j\omega) / I_{SC}(j\omega) = Z_1(j\omega) Z_2(j\omega) / [Z_1(j\omega) - k_1 k_2 Z_2(j\omega)]$

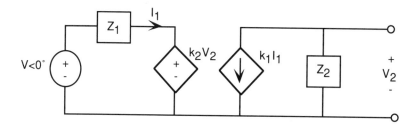

Figure 11.16. Example for Thevenin/Norton equivalence.

11.10   Power in a Sinusoidal Response

11.10.1   Root Mean Square Value

The root mean square (*rms*) value of a sinusoidal signal is often used in power computation. A rms value can be defined for any periodic signal of period T. We begin by asking the question, what should be the output of a constant voltage source V_c applied across a resistor R in order to deliver the same energy to R that is delivered by a sinusoidal voltage source $v(t) = V \sin(\omega t)$ in one period?

Instantaneous power in R is $p(t) = v^2(t)/R = V^2 \sin^2(\omega t) / R$ and energy in one period is $E(t) = \int_0^T p(t)dt$. The energy delivered by the constant source is $T V_c^2/R$. Equating the two expressions and evaluating the integral by using trigonometry we get

$$V_c = V / \sqrt{2} \qquad\qquad (11.10.1)$$

The expression $V / \sqrt{2}$ is called the *root mean square* value of the sinusoid $v(t) = V \sin(\omega t)$ and is often abbreviated as V_{rms}.

11.10.2   Power Expressions

Let $Z(j\omega)$ denote an impedance whose terminal current and voltage are $i(t) = I \sin(\omega t)$ and $v(t) = V \sin(\omega t + \theta)$ respectively. The instantaneous power in the impedance is

$$p(t) = V I \sin(\omega t) \sin(\omega t + \theta)$$
$$= (V I / 2) [\cos(\theta) - \cos(\theta)\cos(2\omega t) + \sin(\theta) \sin(2\omega t)] \qquad (11.10.2)$$

Since the average values of $\sin(2\omega t)$ and $\cos(2\omega t)$ are zeros over a complete period, the average power P_{av} is given by

$$P_{av} = (V I / 2) \cos(\theta)$$
$$= V_{rms} I_{rms} \cos(\theta) \qquad\qquad (11.10.3)$$

Cos(θ) is known as the *power factor* of $\mathbf{Z}(j\omega)$. The real $\mathbf{p}_{real}(t)$ and the reactive $\mathbf{p}_{reactive}(t)$ in $\mathbf{Z}(j\omega)$ are defined as

$$\mathbf{p}_{real}(t) = (\mathbf{V}\,\mathbf{I}\,/\,2)\,[\cos(\theta) - \cos(\theta)\,\cos(2\omega t)], \qquad (11.10.4)$$

$$\mathbf{p}_{reactive}(t) = (\mathbf{V}\,\mathbf{I}\,/\,2)\,\sin(\theta)\,\sin(2\omega t), \qquad (11.10.5)$$

so that

$$\mathbf{p}(t) = \mathbf{p}_{real}(t) + \mathbf{p}_{reactive}(t). \qquad (11.10.6)$$

If $\mathbf{Z}(j\omega)$ has zero reactance i.e., $\theta = 0$ and power factor = 1, then

$$\mathbf{p}(t) = \mathbf{p}_{real}(t) = (\mathbf{V}\,\mathbf{I}\,/\,2)\,[1 - \cos(2\omega t)]$$

and the average power becomes $\mathbf{P}_{av} = (\mathbf{V}\,\mathbf{I}\,/\,2) = \mathbf{V}_{rms}\,\mathbf{I}_{rms}$. On the other hand if $\mathbf{Z}(j\omega)$ is a pure reactance i.e., $\theta = \pm\,90°$ and power factor = 0, then

$$\mathbf{p}(t) = \mathbf{p}_{reactive}(t) = (\mathbf{V}\,\mathbf{I}\,/\,2)\,\sin(2\omega t)$$

and the average power becomes $\mathbf{P}_{av} = 0$. The change in sign of the instantaneous reactive power describes the oscillation of energy between the source and the reactive elements in $\mathbf{Z}(j\omega)$.

11.11   Summary

- The angular frequency ω of a sinusoidal response of a linear circuit is the same as that of the independent sinusoidal source in the circuit.

- The principle of superposition must be used if multiple independent sources of *different* angular frequencies are present.

- Phasors are useful in sinusoidal analysis. All independent sources of the same angular frequency in the circuit are replaced by their phasors. $\mathbf{R}$, $\mathbf{C}$, and $\mathbf{L}$, are represented by their impedances. The phasor of the sinusoidal response is obtained by any one of the techniques used in resistive circuit analysis.

- Instantaneous power in a sinusoidal circuit is the sum of $p_{real}(t)$ and $p_{reactive}(t)$. Real power is a measure of the rate of transfer of energy from electrical to other forms. The reactive power measures the rate of oscillation of electrical energy between a source and the reactive elements in a circuit.

- Average power over one period is the product of V_{rms}, I_{rms} and the power factor $\cos(\theta)$. In a pure reactance, average power is zero.

11.12 Problems

11.1.   Given R_2 and C_2 find the expressions for R_1 and C_1 such that the series circuit in Figure 11.12 (a) has the same impedance as the parallel circuit.

11.2.   Given R_1 and C_1 find the expressions for R_2 and C_2 such that the series circuit in Figure 11.12 (a) has the same impedance as the parallel circuit.

11.3.   Given R_2 and L_2 find the expressions for R_1 and L_1 such that the series circuit in Figure 11.12 (b) has the same impedance as the parallel circuit.

11.4.   Given R_1 and L_1 find the expressions for R_2 and L_2 such that the series circuit in Figure 11.12 (b) has the same impedance as the parallel circuit.

11.5.   Find expressions for the magnitude and the phase of the phasor V_2 in the circuit shown in 11.12 (c). Sketch these expressions as functions of R. Simulate the circuit for suitable parameter values using SPICE and obtain plots for the expressions sketched above.

11.6.   Find an expression for C in the circuit shown in Figure 11.12 (d) so that the power factor of the equivalent impedance is one i.e., the reactive part of the impendence is zero.

11.7. Using node equations find an expression for the voltage phasor $\mathbf{V_R}$ in the circuit shown in Figure 11.12 (e). Validate this expression.

11.8. Using the Thevenin or the Norton equivalent circuit find an expression for the current phasor $\mathbf{I}$ in the circuit shown in Figure 11.12 (f).

11.9. Find the Norton equivalent of the circuit shown in Figure 11.12 (g).

11.10. Using the principle of superposition find an expression for the current phasor $\mathbf{I_R}$ in the circuit shown in Figure 11.12 (h). Validate this expression.

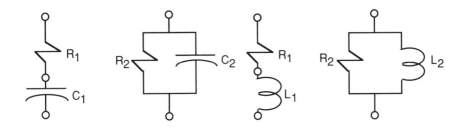

Figure 11.12 (a).   **Figure 11.12 (b).**

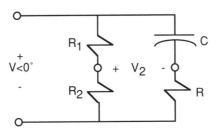

Figure 11.12 (c).

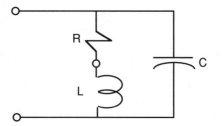

Figure 11.12 (d).

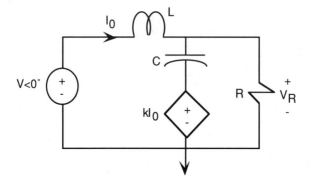

Figure 11.12 (e).

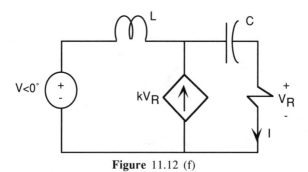

Figure 11.12 (f)

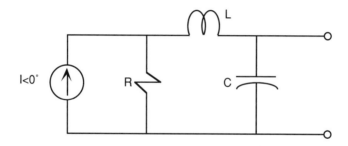

Figure 11.12 (g).

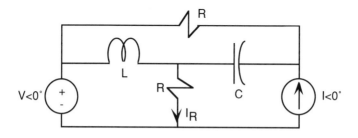

Figure 11.12 (h).

12

FREQUENCY RESPONSE

Sinusoidal analysis provides us with the response of a circuit for a constant angular frequency ω of the input sinusoid. The response is found as a phasor with a magnitude and a phase, both of which are functions of ω as well as other circuit parameters. The *frequency response* of a circuit is defined as the magnitude and the phase of its sinusoidal response as functions of input frequency ω when the input phasor is $1/\underline{0}°$. The magnitude and the phase as functions of ω are called the *magnitude response* and the *phase response* respectively.

If the response of a circuit is comparatively high at certain values of ω than at all the other nearby values of ω, then the circuit exhibits frequency selectivity. Input signals of certain frequencies appear at the output of the circuit in much greater strength than others of different nearby frequencies. This phenomena is called *resonance*.

The frequency response of a circuit exhibits its frequency discrimination characteristics. If the response magnitude is high at some values of ω and low at others, the circuit will respond more to certain values of ω than others in the input. Hence, input signals considered undesirable can be supressed at the output of a circuit if its magnitude response is low at the frequencies of those undesirable input signals. This is the basic concept behind *filtering*. In this chapter we shall discuss resonance and filtering as extensions of simple sinusoidal analysis.

12.1 Series and Parallel Resonance

A simple linear circuit of **R**, **L**, and **C** is made to resonate by taking advantage of the fact that the reactances $j\omega L$ and $-j/\omega C$ change with ω. The reactance of a combination of a **L** and a **C** also changes with ω and can cause the overall impedance of a circuit to change drastically with ω. For example, if a **L** and a **C** are in series, then at $\omega^2 = 1/LC$, the two reactances will cancel each other and the reactance of the overall series combination will become zero. If they are connected in parallel, then their overall reactance will in principle be infinite. Thus simple voltage and current dividers made of **R, L**, and **C** will show remarkable change in response characteristics with ω.

Consider a series connection of **R, L**, and **C** as shown in Figure 12.1 with the input voltage phasor given by $1/\underline{0}°$. Let the circuit response be the terminal voltage of **R**. By voltage division, the voltage drop across **R** is given by

$$V_R = R / [R + j\omega L - j/\omega C]$$
$$= j\omega RC / [(1 - \omega^2 LC) + j\omega RC] \qquad (12.1.1)$$

At $\omega = 0$ the reactance of the capacitor is infinite and the output voltage V_R is zero. Also as ω approaches infinity, the reactance of the inductor approaches infinity and V_R approaches zero. However, at $\omega^2 = 1/LC$, $V_R = 1$, which is its maximum possible value. Figure 12.2 shows the magnitude of V_R as a function of ω for three different series **RLC** circuits.

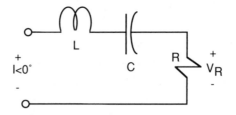

Figure 12.1.

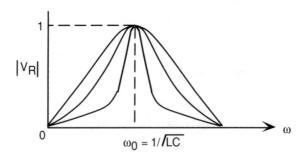

Figure 12.2.

From Figure 12.2 it is clear that at least in some cases there is a rapid increase in the value of V_R as ω approaches $\omega^2 = 1/LC$, and this phenomenon is called *series resonance*. The *resonance frequency* is defined as ω_0 where $\omega_0^2 = 1/LC$. At the resonance frequency, the inductive reactance j $\omega_0 L$ and the capacitive reactance - j $1/\omega_0 C$ add up to zero, and the circuit impedance becomes purely resistive. For $\omega < \omega_0$, the circuit reactance is capacitive and for $\omega > \omega_0$ it is inductive.

For a parallel R, L, and C circuit shown in Figure 12.3 let the input current phasor be $1/\underline{0}°$ and the output response be the terminal current I_R in R. The parallel connection of L and C has the impedance $Z(j\omega) = j\omega L / (1 - \omega^2 LC)$. By current division, the terminal current of R is

$$I_R = Z(j\omega) / [\, R + Z(j\omega) \,]$$
$$= j\omega L / [\, R(1 - \omega^2 LC) + j\omega L \,] \qquad (12.1.2)$$

We see that as a function of ω, the terminal current of R in case of a parallel R, L, and C circuit behaves in a similar manner as the terminal voltage of R in a series R, L, and C circuit. It resonates at $\omega^2 = 1/LC$ and this phenomenon is called *parallel resonance*. The *resonance frequency* is defined as in the case of series resonance.

At the resonance frequency, the inductive reactance j $\omega_0 L$ and the capacitive reactance - j $1/\omega_0 C$ combine in parallel to become infinity. The circuit admittance becomes purely conductive. For $\omega < \omega_0$, the circuit susceptance is inductive and for $\omega > \omega_0$ it is capacitive.

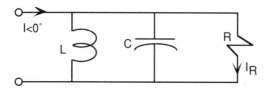

Figure 12.3. Parallel R, L, and C circuit.

12.2 Quality of Resonance

The quality of resonance demonstrated by a circuit is measured by the sharpness in the increase of the circuit response as ω approaches resonance frequency ω_0. In case of a series R, L, and C circuit, the voltage drop across R from (12.1.1) is given by

$$V_R = j\omega RC \,/\, [(1 - \omega^2 LC) + j\omega RC]$$

which can be rewritten as

$$V_R = j \,/\, [j + Q_S \,(\omega_0/\omega - \omega/\omega_0)] \qquad (12.2.1)$$

where $Q_S = \omega_0 L/R$ is called the *quality factor* of the series circuit. We see from (12.2.1) that when ω is close to ω_0, the magnitude of V_R decreases with increasing values of Q_S. Thus a high value of Q_S will keep the magnitude of V_R low when w is in the neighborhood of ω_0. Of course at $\omega = \omega_0$, the magnitude of V_R becomes one. Hence, a high value of Q_S produces a sharp increase in the magnitude of V_R near ω_0 and consequently a sharp resonance.

For a parallel R, L, and C circuit, I_R is given by the same expression as in (12.2.1) except that Q_S is replaced by $Q_P = \omega_0 RC$. In this case Q_P measures the sharpness of resonance and once again a high value of Q_P corresponds to a sharp resonance.

12.3 Bandwidth

The power in the resistor R is proportional to $|V_R|^2$ or $|I_R|^2$ and reaches its maximum value at $\omega = \omega_0$. The *half-power* frequencies ω_1 and ω_2 shown in Figure 12.4, are values of ω where the power in the resistor is half its maximum value. *Bandwidth* **BW** is defined to be $\omega_2 - \omega_1$.

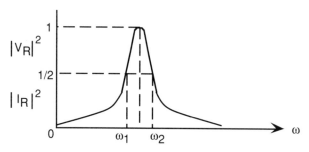

Figure 12.4. *Half-power* frequencies ω_1 and ω_2.

From (12.2.1) we see that at $\omega = \omega_1$ or $\omega = \omega_2$, $Q_S (\omega_0/\omega - \omega/\omega_0) = \pm 1$. This is a quadratic equation in ω and considering the $\pm$ sign on 1 we have two equations. Solving we get four different values of ω and we retain only the positive values. These are

$$\omega_2 = \omega_0/2Q_S + [(\omega_0/2Q_S)^2 + \omega_0^2]^{1/2}, \qquad (12.3.1)$$

$$\omega_1 = - \omega_0/2Q_S + [(\omega_0/2Q_S)^2 + \omega_0^2]^{1/2}. \qquad (12.3.2)$$

The bandwidth **BW** is obviously

$$\mathbf{BW} = \omega_0 / Q_S. \qquad (12.3.3)$$

A similar expression for the bandwidth of the parallel **R, L,** and **C** circuit can be obtained where Q_S is replaced by Q_P. The higher the quality factor Q_S or Q_P, the narrower the bandwidth **BW**, and the sharper the resonance.

12.4 Simple Filters

Filters are useful for noise elimination in signal processing and can be quite complex circuits. However, the basic idea behind filtering a noisy signal can be illustrated by means of simple circuits. An ideal filter allows input signals of certain specified frequencies to pass through to its output. Input signals of other frequencies are totally supressed. The output response vs. input frequency of three ideal filters are shown in Figure 12.5.

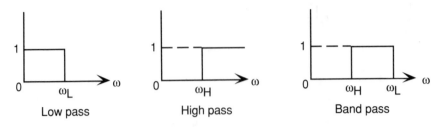

Figure 12.5. Three ideal filters.

For a *low pass* ideal filter, the output response beyond a certain cutoff frequency is zero. Thus only low frequency input signals, characterized by the cutoff frequency, ever produce a nonzero output response. The reasons for naming the other ideal filters *high pass* and *band pass* should now be obvious from the Figure 12.5.

A simple (non-ideal) low pass filter can be realized by a RC voltage divider shown in Figure 12.6. At low frequencies the reactance of the capacitor is high. Most of the input voltage appears across the capacitor. However, at high frequencies, the capacitor becomes a virtual short circuit and the output voltage approaches zero. The output voltage across the capacitor is given by

$$V_C = 1 / (1 + jwRC), \qquad\qquad (12.4.1)$$

and a sketch of $|V_C|^2$ vs. ω is shown in Figure 12.7. At $\omega = 0$, $|V_C| = 1$, and as ω approaches infinity $|V_C|$ approaches zero. At $\omega = 1/RC$, $|V_C|^2 = 1/2$, and this half-power frequency is defined as the *cutoff* frequency of this low pass filter.

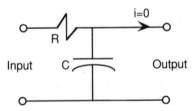

Figure 12.6. Simple low pass filter.

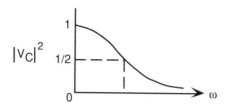

Figure   12.7.

The voltage divider of Figure 12.8 is a (non-ideal) high pass filter. At low frequencies the reactance of the capacitor is high and the voltage drop across the resistor **R** (the output voltage) is very low. At high frequencies the capacitor becomes a virtual short circuit and the output voltage approaches the input voltage in magnitude. The output voltage across the resistor is given by

$$V_R = j\omega RC / (1 + j\omega RC), \tag{12.4.2}$$

and a sketch of $|V_R|^2$ vs. ω is shown in Figure 12.9. At $\omega = 0$, $|V_R| = 0$, and as ω approaches infinity $|V_R|$ approaches one. Again at $\omega = 1/RC$, $|V_R|^2 = 1/2$, and this half-power frequency is defined as the cutoff frequency of this high pass filter.

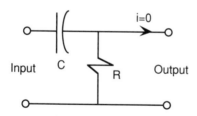

Figure 12.8. Simple high pass filter.

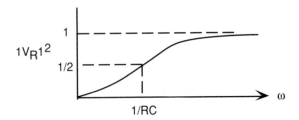

Figure 12.9.

A simple (non-ideal) band pass filter can be obtained by joining a low pass filter in series with a high pass filter as shown in Figure 12.10. Since both the low and the high input frequencies are suppressed from the output voltage, only a band of medium input frequencies are transmitted. Note that the expression of the output voltage is not simply the product of (12.4.1) and (12.4.2) and must be derived by circuit analysis. The half power cutoff frequencies are also given by more complex expressions in this case.

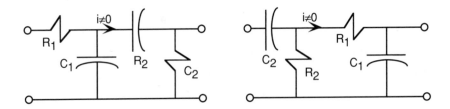

Figure 12.10. A simple (non-ideal) band pass filter.

12.5 SPICE Simulations

Resonant and filter circuits are analyzed by means of phasors introduced in the previous chapter. Hence, SPICE simulation of these circuits are similar to those demonstrated there.

Example 12.5.1 Resonance

Figure 12.11 shows the circuit file of a simple parallel **R**, **L**, and **C** circuit followed by the plot of the magnitude of $|\mathbf{I_R}|^2$ vs. input frequency. The occurrence of resonance is obvious from the plot shown in Figure 12.12. It is not hard to verify the values of the resonance frequency, the quality factor, and the bandwidth from the given circuit parameter values and the plot. However, SPICE simulation can be used to verify resonance in more complex but practical circuits.

```
reson1.cir   parallel resonant circuit
*  input sinusoidal source
*  name   nodes    ac    amplitude   phase
is        0   1    ac        1         0

r1    1    2     100
vd1   2    0     dc     0V

c1    1    0     1M

l1    1    0     1M

.ac   lin   100   155   165
.probe
.options   nopage
.end
```

Figure 12.11. Circuit file of a simple parallel **R**, **L**, and **C** circuit.

Most practical inductors have internal resistances that are ignored to simplify analysis. How does such internal resistance of an inductor affect parallel resonance? The obvious approach is to assume that the internal resistance is in series with the inductor and thus create an approximate circuit such as the one shown in Figure 12.13.

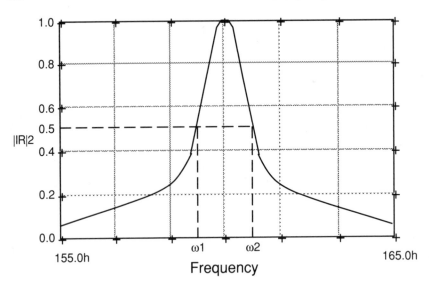

Figure   12.12.

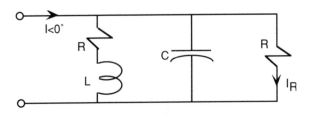

Figure   12.13.

Figure 12.14 shows the circuit file of a circuit with a resistor in series with the inductor and the plot of the magnitude of $|I_R|^2$ vs. input frequency is shown in Figure 12.15. First we note that the magnitude of I_R is not zero as $\omega = 0$. This is because although the inductor is a short circuit at $\omega = 0$, the resistance in series with it is not zero. Hence, we have a current divider instead of a short circuit across the resistor **R** (Figure 12.13). The higher the internal resistance of the inductor, the larger the magnitude of I_R at $\omega = 0$ due to the operation of this current divider.

```
reson2.cir  approximate parallel resonant circuit
* input sinusoidal source
* name  nodes   ac    amplitude    phase
is       0   1   ac        1           0

r1   1     2     100
vd1  2     0     dc    0V

c1   1     0     1M
* internal resistance of inductor r2
r2   1     3     1
l1   3     0     1M

.ac   lin   200  10   300
.probe
.options   nopage
.end
```

Figure 12.14. Circuit file of a circuit with a resistor in series with the inductor.

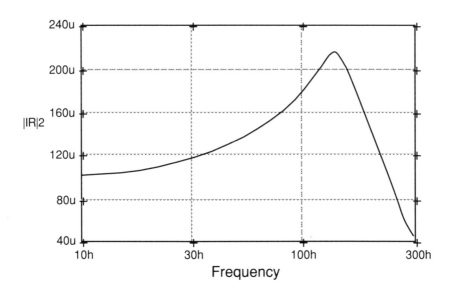

Figure 12.15.

The expression of ω at which the admittance of the parallel combination of **C** and **L** in series with **r** (Figure 12.13) becomes zero is no longer given by $\omega_0^2 = 1/LC$. Simple analysis shows that this expression is

$$\omega_0^2 = [(1/LC) - (r/L)^2]. \tag{12.5.1}$$

Clearly from (12.5.1), as the value of the internal resistance of the inductor (**r**) increases, the value of ω_0 decreases from its ideal value of $1/LC$. However, the magnitude of I_R does not attain its maximum at the value of ω given by (12.5.1). The expression for the maximizing value of ω is much more complex than before.

Example 12.5.2 Filters

Simple **RC** low, high and band pass filters can be easily simulated in SPICE. Figures 12.16 and 12.17 show the circuit files for simple **RC** low and high pass filters respectively (Figures 12.6 and 12.7). The squares of the magnitudes of the output voltages as functions of ω are also shown in Figures 12.18 and 12.19. Half-power cutoff frequencies can be easily read from these plots and compared against theoretical values given by the circuit parameters.

```
filt1.cir   lowpass rc filter
* input sinusoidal source
* name   nodes     ac    amplitude    phase
vs         0    1    ac        1           0

r1    1     2       10
c1    2     0       0.1M

.ac   lin   100   1   500
.probe
.options   nopage
.end
```

Figure 12.16. Circuit file for simple **RC** low pass filter.

```
filt2.cir  highpass rc filter
* input sinusoidal source
* name   nodes   ac   amplitude   phase
vs       0   1   ac       1         0

r1   1    2     10
c1   2    0     0.5M
* ouput voltage across r1, nodes (1,2)

.ac  lin  100  1  500
.probe
.options  nopage
.end
```

Figure 12.17. Circuit file for simple **RC** high pass filter.

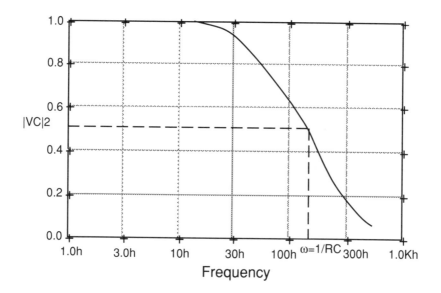

Figure 12.18.

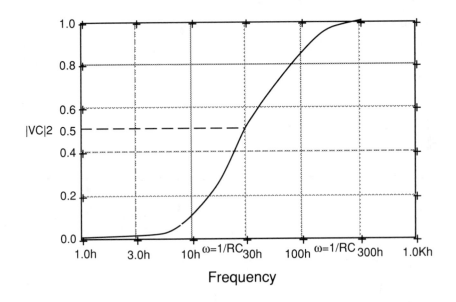

Figure   12.19.

SPICE simulation allows us to easily explore possible structures of circuits which may be more complex and time-consuming for analysis. As a simple example let us consider a band pass filter constructed out of the low and the high pass filters simulated already. The low and the high pass filters are connected in series. There are two possible ways and these are shown in Figure 12.10. Are the voltage outputs identical for these two circuits?

Figure 12.20 shows the circuit file for simulating the two band pass filters. Their output voltages as functions of ω are shown in Figure 12.21. Obviously these output voltages are different. Let us explore what these differences are and why they are there in the first place.

The graphs in Figure 12.21 show that the maximum of the output voltages of the two bandpass filters are different. Furthermore, their cutoff frequencies and consequently their bandwidths are different. Note that the output responses of the low and the high pass filters were obtained by simple voltage division, i.e., with no load impedence on the output side. When these circuits are connected in series, the circuit on the output side draws current from and hence *loads* the citcuit on the input side. Hence, the overall circuit behavior changes and can only be ascertained by a new analysis of the series combination. Since the low and the high pass filters are different circuits, their loading effects on each other are also different. Hence, the two series combinations produce such different bandpass filters.

```
fil5.cir   two alternate bandpass rc filters
*  input  sinusoidal source
*  name   nodes     ac    amplitude    phase
vs           0   1   ac        1           0

r1    1      2      10
c1    2      0      0.1M

r2    3      0      10
c2    2      3      0.5M
*<<<<<<<<<<<<<<<<<<<<<<<<<<<<<<<<<<<<<<<<
r3    4      0      10
c3    1      4      0.5M

r4    4      5      10
c4    5      0      0.1M

.ac   lin   100   1   500
.probe
.options   nopage
.end
```

Figure 12.20. Circuit file for simulating the two band pass filters.

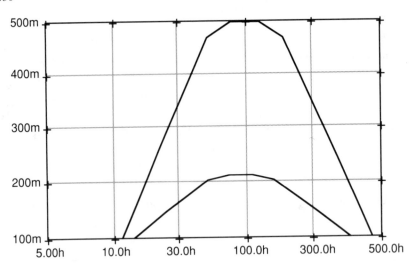

Figure 12.21.

12.6 Summary

- Variations in the output voltage or current with changes in the frequency of an input sinusoid is defined as the *frequency response* of a circuit. An expression for the frequency response of a circuit can be found by phasor analysis assuming that the input phasor is $1/\underline{0}°$. Each phasor response provides a *magnitude response* and a *phase response*.

- The frequency response of a circuit exhibits its frequency discrimination characteristics. The magnitude response can be used to describe circuits which resonate at a particular input frequency or filter a range of input frequencies.

- The resonance phenomenon is characterized by its *resonance frequency* and *bandwidth*. For simple series or parallel **R, L,** and **C** circuits, expressions for these features can be obtained by circuit analysis. More complex circuits can be studied by SPICE simulations.

- Filters are characterized by their *half-power* frequencies or *cutoff* frequencies. Simple *low* and *high pass* filters can be obtained by voltage dividers using resistors and capacitors. *Band pass* filters can be obtained by series connections of a low and a high pass filter. Because of the effect of one circuit *loading* another, such series connections have to be reanalyzed.

12.7 Problems

12.1.   Derive (12.2.1).

12.2.   Show that for a parallel **R**, **L**, and **C** circuit $\mathbf{I_R}$ is given by the same expression as in (12.2.1) except that $\mathbf{Q_S}$ is replaced by $\mathbf{Q_P} = \omega_o \mathbf{RC}$. Also show that $\mathbf{Q_S} = 1 / \mathbf{Q_P}$.

12.3.   Derive (12.3.1), (12.3.2), and (12.3.3).

12.4.   Consider the filter shown in Figure 12.22. Find an expression for its cutoff frequency and sketch its amplitude response.

12.5.   Consider the filter shown in Figure 12.23. Find an expression for its cutoff frequency and sketch its amplitude response.

12.6.   Using SPICE simulate the two filters analyzed in Problems 12.4 and 12.5.

12.7.   Design two band pass filters using the two filters analyzed in Problems 12.4 and 12.5. Use SPICE to simulate these two band pass filters and compare their magnitude responses.

12.8.   Derive (12.5.1).

12.9.   Using SPICE simulate the circuit shown in Figure 12.24 and plot its magnitude response. What sort of discrimination of input sinusoid frequency does this circuit show?

12.10. Using SPICE simulate the circuit shown in Figure 12.25 and plot its magnitude
response. What sort of discrimination of input sinusoid frequency does this
circuit show?

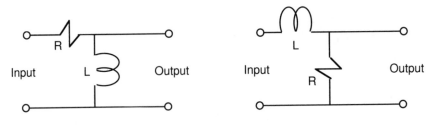

Figure 12.22.   Figure 12.23.

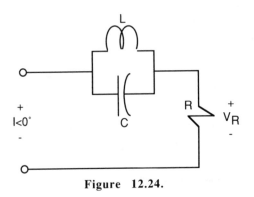

Figure 12.24.

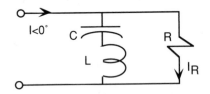

Figure 12.25.

Appendix A

ALGEBRA OF COMPLEX NUMBERS

A.1 The Need for Complex Numbers

Complex numbers are absolutely essential for solving many engineering problems. For example, solutions to many problems require the computation of the roots of a quadratic polynomial such as

$$a\,x^2 + b\,x + c = 0, \ (a \neq 0) \tag{A.1.1}$$

To find these roots of (A.1.1), we divide both sides of (A.1.1) by a and completing the square rewrite it as:

$$(x + b/2a)^2 - (b/2a)^2 + (c/a) = 0. \tag{A.1.2}$$

Solving for x from (A.1.2) we get

$$x = (-\,b \pm (b^2 - 4ac)^{1/2}\,) \,/\, 2a. \tag{A.1.3}$$

The general expression (A.1.3) for the roots of a quadratic polynomial is easy enough to derive. However, we run into serious problems as soon as we apply it to find the roots in specific cases. On occasions it does not work at all. For example cosider the roots of

$$x^2 + 4\,x + 5 = 0. \tag{A.1.4}$$

According to (A.1.3), the roots are

$$x = -2 \pm (-1)^{1/2}. \qquad (A.1.5)$$

We can not compute the square root of a negative number since the squares of all real numbers are positive. However, in engineering problems the roots usually represent some physical quantities. Hence, we know that they do exist and must be computable.

So we realize that we must be running into some mathematical limitations and must extend our notion of numbers beyond the real numbers. Complex numbers are such an extension. First we introduce a new symbol $j = (-1)^{1/2}$. In terms of this new symbol we can write the roots of (A.1.4) as

$$x = -2 \pm j. \qquad (A.1.6)$$

The values of x in (A.1.6) are called *complex* numbers. In general, a complex number d appears as

$$d = -e \pm jf, \qquad (A.1.7)$$

where e and f are real numbers and $j = (-1)^{1/2}$. The 'e' and the 'f' are called the *real* and the *imaginary* parts of 'd' respectively and often written as $e = \text{Re}[d]$, $f = \text{Im}[d]$.

A.2 An Algebra of Complex Numbers

So far we have not done much other than introduce a new symbol 'j' and call certain expressions complex numbers. To understand complex numbers better and to use them freely we must create an algebra of complex numbers. The first step in this process is to figure out how to do arithmetic operations with complex numbers. Since the complex numbers come out of our analysis of the roots of a quadratic polynomial, it is natural to observe how they interact with each other in such a polynomial and derive the rules for arithmetic operations from these observations. If r_1 and r_2 are the roots of a quadratic polynomial, then according to the fundamental theorem of algebra we can write

$$a x^2 + b x + c = (x - r1)(x - r2). \qquad (A.2.1)$$

In case of (A.1.4) the corresponding identity is

$$x^2 + 4 x + 5 = (x + 2 + j) (x + 2 - j)$$
$$= x^2 + x (2-j+2+j) + (4+2j-2j-j^2). \qquad (A.2.2)$$

Comparing the coefficients from both sides of (A.2.2) we conclude

$$4 = (2 - j + 2 + j), \qquad (A.2.3)$$

and

$$5 = (2 + j) (2 - j)$$
$$= (4 - j^2) + j (2 - 2). \qquad (A.2.4)$$

From these observations, the rules for the addition, the subtraction, and the multiplication operations can be inductively derived as:

$$(a + jb) \pm (c + jd) = (a \pm c) + j (b \pm d), \qquad (A.2.5)$$

and

$$(a + jb) (c + jd) = (ac - bd) + j (ad + bc). \qquad (A.2.6)$$

Now that we know how to multiply complex numbers we can figure out a method of dividing one complex number by another. The derivation of this rule is given below.

$$(a + jb) / (c + jd) = [(a + jb) / (c + jd)] [(c - jd) / (c - jd)]$$
$$= [(a + jb) (c - jd)] / (c^2 + d^2). \qquad (A.2.7)$$

The numerator on the right hand side (RHS) of (A.2.7) is a product of two complex numbers which can be computed by (A.2.6) and written in terms of its real and the imaginary parts. The denominator on the RHS of (A.2.7) is a real number and hence, the division of the real and the imaginary parts of the numerator is by a real number only.

A.3   Representation of Complex Numbers

Real numbers can be represented as points along a line. In a similar fashion, complex numbers can be represented as points on a plane called the *complex plane*. The real and the imaginary parts of a complex number are real numbers and are measured off along the horizontal and the vertical axes respectively. Figure A.3 shows such a representation of complex numbers on a plane. This form of representation is called the *rectangular form*.

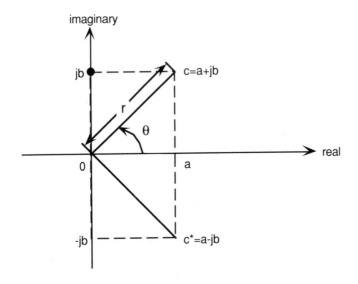

Figure A.3. Representation of complex numbers on a plane.

From Figure A.3 we see that a point c = a + jb on the complex plane can also be denoted by its distance r from the origin and the angle θ measured from the positive direction of the horizontal axis. Obviously

$$r^2 = a^2 + b^2, \tag{A.3.1}$$

and

$$\theta = \arctan(b/a). \qquad (A.3.2)$$

Also

$$a = r \cos\theta), \qquad (A.3.3)$$

and

$$b = r \sin(\theta). \qquad (A.3.4)$$

This is called a *polar form* representation of a complex number and written as $r \,/\_\theta$.

For every complex number $c = a + jb$, the related number $c^* = a - jb$ is called its *complex conjugate* (see Figure A.3). Using any complex number c and its complex conjugate c^* we can write

$$a = \text{Re}[c] = (c + c^*) / 2, \qquad (A.3.5)$$
$$b = \text{Im}[c] = (c - c^*) / 2, \qquad (A.3.6)$$
$$r^2 = c\,c^*, \qquad (A.3.7)$$

and

$$\theta = \arctan((c - c^*) / (c + c^*)). \qquad (A.3.8)$$

A.4 Complex Variables and Functions

Given that $\mathbf{x}$ and $\mathbf{y}$ are real variables, $\mathbf{z} = \mathbf{x} + j\mathbf{y}$ is a complex variable. The theory of functions of a complex variable $f(\mathbf{z})$ is rather extensive. Fortunately, for our purpose we need some rudimentary knowledge. The complex exponential function $\exp(\mathbf{z})$ should satisfy the exponential relation $\exp(\mathbf{z}_1 + \mathbf{z}_2) = \exp(\mathbf{z}_1) \bullet \exp(\mathbf{z}_2)$. Formally we define the complex $\exp(\mathbf{z})$ by the infinite series used to define the real exponential $\exp(\mathbf{x})$.

$$\exp(\mathbf{z}) = 1 + \mathbf{z} + \mathbf{z}^2/2! + \mathbf{z}^3/3! + \mathbf{z}^4/4! + \dots \qquad (A.4.1)$$

Now we substitute $z = j\theta$ in (A.4.1) and collect the real and the imaginary terms on the RHS.

$$\exp(j\theta) = [1 - \theta^2/2! + \theta^4/4! + ...] + j [\theta - \theta^3/3! + \theta^5/5! + ...] \qquad (A.4.2)$$

The series expressions on the RHS of (A.4.2) are the series expansions of the cosine and the sine functions respectively. Replacing these series expansions by their cosine and sine forms respectively we have the famous and extremely useful *Euler identity*:

$$\exp(j\theta) = \cos(\theta) + j \sin(\theta). \qquad (A.4.3)$$

Differentiating both sides of (A.4.3) with respect to θ and using Euler identity again we get

$$d\exp(j\theta)/d\theta = j \exp(j\theta), \qquad (A.4.4)$$

and integrating both sides of (A.4.3) with respect to θ and using Euler identity we have

$$\int \exp(j\theta) \, d\theta = \exp(j\theta) / j. \qquad (A.4.5)$$

Let $c = a + jb = r \cos(\theta) + j r \sin(\theta)$ be any complex number. Then $c = r (\cos(\theta) + j \sin(\theta)) = r \exp(j\theta)$ by Euler identity. This last expression is a new representation of a complex number called its *phasor form*. The multiplication and the division operations are particularly easy to carry out when complex numbers are in their phasor forms. The equations (A.2.6) and (A.2.7) can be rewritten as:

$$r_1 \exp(\theta_1) \, r_2 \exp(\theta_2) = r_1 \, r_2 \exp(\theta_1 + \theta_2), \qquad (A.4.6)$$

and

$$r_1 \exp(\theta_1) / r_2 \exp(\theta_2) = (r_1 / r_2) \exp(\theta_1 - \theta_2). \qquad (A.4.7)$$

Index